Penguin Reference Books

The Penguin Book of Mathematical and Statistical Tables

David Nelson was educated at Calday Grange Grammar School, Cheshire, and won an open scholarship to Christ's College, Cambridge. After postgraduate studies in mathematical logic at Cambridge and at Bristol University, he entered the teaching profession. He has published papers in mathematical journals and has been Penguin's Advisory Editor for mathematics and statistics since 1971.

The Penguin Book of Mathematical and Statistical Tables

PENGUIN BOOKS

R. D. Nelson

Penguin Books Ltd, Harmondsworth,
Middlesex, England
Penguin Books, 625 Madison Avenue,
New York, New York 10022, U.S.A.
Penguin Books Australia Ltd, Ringwood,
Victoria, Australia
Penguin Books Canada Ltd, 2801 John Street,
Markham, Ontario, Canada L3R 1B4
Penguin Books (N.Z.) Ltd, 182–190 Wairau Road,
Auckland 10, New Zealand

First published 1980

Made and printed in Great Britain by Butler & Tanner Ltd, Frome, Somerset
Set in Ehrhardt

Contents

Preface

It is hoped that this collection will prove sufficiently comprehensive to be suitable for the teaching of mathematics and statistics in schools, polytechnics and universities, and, at the same time, will meet the needs of many of the users of statistics in research and industry.

Tables are given for all the basic statistical functions and parametric tests. In addition, there are tables for four non-parametric tests, namely the sign test, Wilcoxon's signed rank test, and the rank correlation methods of Spearman and of Kendall.

As there is no universal convention, some degree of internal consistency in tabulating statistical functions has been attempted in this volume. In particular, all the tables of percentage points and critical values are given for single-tailed and one-sided tests of significance. Short notes at the foot of each statistical table explain what is actually tabulated and it is anticipated that, once the reader is familiar with the book, this uniformity will be found an advantage.

I am indebted to the editors of the *Annals of Mathematical Statistics* for permission to calculate Table 29 from the material in the paper 'Distribution of sums of squares of rank differences for small numbers of individuals', by E. G. Olds (1938).

Tables 27, 30 and 35 have been reprinted and adapted from those in other books, and I am most grateful to the authors and publishers of the following works for permission to use their material:

Statistical Tables for Biological, Agricultural and Medical Research, by the late Sir Ronald A. Fisher and Dr Frank Yates, published by Longmans (Table VII); *Rank Correlation Methods*, by Sir Maurice Kendall, published by Griffin (Table 1); and *Critical Values and Probability Levels for the Wilcoxon Rank Sum Test and the Wilcoxon Signed Rank Test*, by the late F. Wilcoxon, S. K. Katti and Roberta A. Wilcox, published by American Cyanamid Company and The Florida State University (Table II).

Notes on Interpolation

The mathematical tables in this book give the values $y_1, y_2, y_3, \ldots$ of a function for a number of evenly spaced values $x_1, x_2, x_3, \ldots$ of an independent variable x. The tabular values are usually displayed in rows and the tabular interval h between successive values of x is constant. Thus in Table 1 there are 90 rows of tabular values and $h = 0{\cdot}1$.

When the value of a function is required for an intermediate value of x, lying between x_1 and x_2 for example, then some method of interpolation must be used.

Linear interpolation. This amounts to taking the graph of the function to be the straight line joining the points (x_1, y_1) and (x_2, y_2). Let the difference $y_2 - y_1$ between the successive tabular values be denoted Δy_1. Then the value of the function at $x = x_1 + ph$, where $0 < p < 1$, is taken to be $y_1 + p\Delta y_1$.

Example: to find sin 26·74.

$$\begin{aligned} \text{Here}\quad y_1 &= 0{\cdot}4493,\ h = 0{\cdot}1, \text{ and } p = 0{\cdot}4. \\ \Delta y_1 &= \sin 26{\cdot}8 - \sin 26{\cdot}7 = 0{\cdot}0016 \\ \text{so}\quad \sin 26{\cdot}74 &= 0{\cdot}4493 + 0{\cdot}4 \times 0{\cdot}0016 \\ &= 0{\cdot}4499 \text{ to 4 decimal places.} \end{aligned}$$

Mean differences. In general, the value of Δy varies very little along a row and adequate accuracy in interpolation can be obtained by using the set of mean differences given at the end of each row. However, at the fringes of some of the tables, Δy varies significantly along the rows and the use of mean differences can lead to errors of 3 units or more in the last figure. An asterisk marks the last (or first) row where such errors arise from the use of mean differences at the beginning (or end) of a table. To obtain reliable results linear or quadratic interpolation should be used in these parts of the tables.

Quadratic interpolation. If the difference between successive values of Δy exceeds 10 units, then linear interpolation ceases to give adequate accuracy and more precise methods such as quadratic interpolation are necessary. The form of quadratic interpolation given here amounts to taking the graph of the function to be the quadratic through the points (x_1, y_1), (x_2, y_2) and (x_3, y_3).

Let the successive differences $y_2 - y_1$ and $y_3 - y_2$ be denoted Δy_1 and Δy_2 respectively. Then $\Delta y_2 - \Delta y_1$ is denoted $\Delta^2 y_1$. With this notation, the value of the function at $x = x_1 + ph$ is taken to be

$$y_1 + p\Delta y_1 + \tfrac{1}{2}\, p(p-1)\Delta^2 y_1.$$

This value differs from that given by linear interpolation by at most $\frac{1}{8}|\Delta^2 y_1|$ units.

Example: to find tan 83·35.

$$\begin{aligned} y_1 &= 8{\cdot}5126,\ h = 0{\cdot}1 \text{ and } p = 0{\cdot}5. \\ \Delta y_1 &= 0{\cdot}1301,\ \Delta y_2 = 0{\cdot}1342 \text{ and } \Delta^2 y_1 = 0{\cdot}0041 \\ \text{so } \tan 83{\cdot}35 &= 8{\cdot}5126 + 0{\cdot}06505 - 0{\cdot}00051 \\ &= 8{\cdot}5771. \end{aligned}$$

Interpolation linear in $\frac{120}{x}$. This is required in Tables 31 and 33, and is performed by regarding y as a linear function of $\frac{120}{x}$.

Example: to find t_p for $v = 100$, $p = 0{\cdot}005$ in Table 31.

$$\begin{aligned} \text{Here } x &= 60,\ 100,\ 120 \text{ and } \tfrac{120}{x} = 2,\ 1{\cdot}2,\ 1. \\ y_1 &= 2{\cdot}660,\ h = -1,\ p = 0{\cdot}8 \text{ and} \\ \Delta y_1 &= -0{\cdot}043 \\ \text{so}\quad t_p &= 2{\cdot}660 - 0{\cdot}8 \times 0{\cdot}043 = 2{\cdot}626. \end{aligned}$$

Table 1. Natural Sines $\sin x°$

$x°$	0°·0 0′	0°·1 6′	0°·2 12′	0°·3 18′	0°·4 24′	0°·5 30′	0°·6 36′	0°·7 42′	0°·8 48′	0°·9 54′	1′	2′	3′	4′	5′
0°	0·0000	0·0017	0·0035	0·0052	0·0070	0·0087	0·0105	0·0122	0·0140	0·0157	3	6	9	12	15
1	0·0175	0·0192	0·0209	0·0227	0·0244	0·0262	0·0279	0·0297	0·0314	0·0332	3	6	9	12	15
2	0·0349	0·0366	0·0384	0·0401	0·0419	0·0436	0·0454	0·0471	0·0488	0·0506	3	6	9	12	15
3	0·0523	0·0541	0·0558	0·0576	0·0593	0·0610	0·0628	0·0645	0·0663	0·0680	3	6	9	12	15
4	0·0698	0·0715	0·0732	0·0750	0·0767	0·0785	0·0802	0·0819	0·0837	0·0854	3	6	9	12	14
5	0·0872	0·0889	0·0906	0·0924	0·0941	0·0958	0·0976	0·0993	0·1011	0·1028	3	6	9	12	14
6	0·1045	0·1063	0·1080	0·1097	0·1115	0·1132	0·1149	0·1167	0·1184	0·1201	3	6	9	12	14
7	0·1219	0·1236	0·1253	0·1271	0·1288	0·1305	0·1323	0·1340	0·1357	0·1374	3	6	9	12	14
8	0·1392	0·1409	0·1426	0·1444	0·1461	0·1478	0·1495	0·1513	0·1530	0·1547	3	6	9	12	14
9	0·1564	0·1582	0·1599	0·1616	0·1633	0·1650	0·1668	0·1685	0·1702	0·1719	3	6	9	11	14
10	0·1736	0·1754	0·1771	0·1788	0·1805	0·1822	0·1840	0·1857	0·1874	0·1891	3	6	9	11	14
11	0·1908	0·1925	0·1942	0·1959	0·1977	0·1994	0·2011	0·2028	0·2045	0·2062	3	6	9	11	14
12	0·2079	0·2096	0·2113	0·2130	0·2147	0·2164	0·2181	0·2198	0·2215	0·2233	3	6	9	11	14
13	0·2250	0·2267	0·2284	0·2300	0·2317	0·2334	0·2351	0·2368	0·2385	0·2402	3	6	8	11	14
14	0·2419	0·2436	0·2453	0·2470	0·2487	0·2504	0·2521	0·2538	0·2554	0·2571	3	6	8	11	14
15	0·2588	0·2605	0·2622	0·2639	0·2656	0·2672	0·2689	0·2706	0·2723	0·2740	3	6	8	11	14
16	0·2756	0·2773	0·2790	0·2807	0·2823	0·2840	0·2857	0·2874	0·2890	0·2907	3	6	8	11	14
17	0·2924	0·2940	0·2957	0·2974	0·2990	0·3007	0·3024	0·3040	0·3057	0·3074	3	6	8	11	14
18	0·3090	0·3107	0·3123	0·3140	0·3156	0·3173	0·3190	0·3206	0·3223	0·3239	3	6	8	11	14
19	0·3256	0·3272	0·3289	0·3305	0·3322	0·3338	0·3355	0·3371	0·3387	0·3404	3	5	8	11	14
20	0·3420	0·3437	0·3453	0·3469	0·3486	0·3502	0·3518	0·3535	0·3551	0·3567	3	5	8	11	14
21	0·3584	0·3600	0·3616	0·3633	0·3649	0·3665	0·3681	0·3697	0·3714	0·3730	3	5	8	11	14
22	0·3746	0·3762	0·3778	0·3795	0·3811	0·3827	0·3843	0·3859	0·3875	0·3891	3	5	8	11	13
23	0·3907	0·3923	0·3939	0·3955	0·3971	0·3987	0·4003	0·4019	0·4035	0·4051	3	5	8	11	13
24	0·4067	0·4083	0·4099	0·4115	0·4131	0·4147	0·4163	0·4179	0·4195	0·4210	3	5	8	11	13
25	0·4226	0·4242	0·4258	0·4274	0·4289	0·4305	0·4321	0·4337	0·4352	0·4368	3	5	8	11	13
26	0·4384	0·4399	0·4415	0·4431	0·4446	0·4462	0·4478	0·4493	0·4509	0·4524	3	5	8	10	13
27	0·4540	0·4555	0·4571	0·4586	0·4602	0·4617	0·4633	0·4648	0·4664	0·4679	3	5	8	10	13
28	0·4695	0·4710	0·4726	0·4741	0·4756	0·4772	0·4787	0·4802	0·4818	0·4833	3	5	8	10	13
29	0·4848	0·4863	0·4879	0·4894	0·4909	0·4924	0·4939	0·4955	0·4970	0·4985	3	5	8	10	13
30	0·5000	0·5015	0·5030	0·5045	0·5060	0·5075	0·5090	0·5105	0·5120	0·5135	3	5	8	10	13
31	0·5150	0·5165	0·5180	0·5195	0·5210	0·5225	0·5240	0·5255	0·5270	0·5284	2	5	7	10	12
32	0·5299	0·5314	0·5329	0·5344	0·5358	0·5373	0·5388	0·5402	0·5417	0·5432	2	5	7	10	12
33	0·5446	0·5461	0·5476	0·5490	0·5505	0·5519	0·5534	0·5548	0·5563	0·5577	2	5	7	10	12
34	0·5592	0·5606	0·5621	0·5635	0·5650	0·5664	0·5678	0·5693	0·5707	0·5721	2	5	7	10	12
35	0·5736	0·5750	0·5764	0·5779	0·5793	0·5807	0·5821	0·5835	0·5850	0·5864	2	5	7	9	12
36	0·5878	0·5892	0·5906	0·5920	0·5934	0·5948	0·5962	0·5976	0·5990	0·6004	2	5	7	9	12
37	0·6018	0·6032	0·6046	0·6060	0·6074	0·6088	0·6101	0·6115	0·6129	0·6143	2	5	7	9	12
38	0·6157	0·6170	0·6184	0·6198	0·6211	0·6225	0·6239	0·6252	0·6266	0·6280	2	5	7	9	11
39	0·6293	0·6307	0·6320	0·6334	0·6347	0·6361	0·6374	0·6388	0·6401	0·6414	2	4	7	9	11
40	0·6428	0·6441	0·6455	0·6468	0·6481	0·6494	0·6508	0·6521	0·6534	0·6547	2	4	7	9	11
41	0·6561	0·6574	0·6587	0·6600	0·6613	0·6626	0·6639	0·6652	0·6665	0·6678	2	4	7	9	11
42	0·6691	0·6704	0·6717	0·6730	0·6743	0·6756	0·6769	0·6782	0·6794	0·6807	2	4	6	9	11
43	0·6820	0·6833	0·6845	0·6858	0·6871	0·6884	0·6896	0·6909	0·6921	0·6934	2	4	6	8	11
44	0·6947	0·6959	0·6972	0·6984	0·6997	0·7009	0·7022	0·7034	0·7046	0·7059	2	4	6	8	10

x°	0°·0 0′	0°·1 6′	0°·2 12′	0°·3 18′	0°·4 24′	0°·5 30′	0°·6 36′	0°·7 42′	0°·8 48′	0°·9 54′	1′	2′	3′	4′	5′
45°	0·7071	0·7083	0·7096	0·7108	0·7120	0·7133	0·7145	0·7157	0·7169	0·7181	2	4	6	8	10
46	0·7193	0·7206	0·7218	0·7230	0·7242	0·7254	0·7266	0·7278	0·7290	0·7302	2	4	6	8	10
47	0·7314	0·7325	0·7337	0·7349	0·7361	0·7373	0·7385	0·7396	0·7408	0·7420	2	4	6	8	10
48	0·7431	0·7443	0·7455	0·7466	0·7478	0·7490	0·7501	0·7513	0·7524	0·7536	2	4	6	8	10
49	0·7547	0·7559	0·7570	0·7581	0·7593	0·7604	0·7615	0·7627	0·7638	0·7649	2	4	6	8	9
50	0·7660	0·7672	0·7683	0·7694	0·7705	0·7716	0·7727	0·7738	0·7749	0·7760	2	4	6	7	9
51	0·7771	0·7782	0·7793	0·7804	0·7815	0·7826	0·7837	0·7848	0·7859	0·7869	2	4	5	7	9
52	0·7880	0·7891	0·7902	0·7912	0·7923	0·7934	0·7944	0·7955	0·7965	0·7976	2	4	5	7	9
53	0·7986	0·7997	0·8007	0·8018	0·8028	0·8039	0·8049	0·8059	0·8070	0·8080	2	3	5	7	9
54	0·8090	0·8100	0·8111	0·8121	0·8131	0·8141	0·8151	0·8161	0·8171	0·8181	2	3	5	7	8
55	0·8192	0·8202	0·8211	0·8221	0·8231	0·8241	0·8251	0·8261	0·8271	0·8281	2	3	5	7	8
56	0·8290	0·8300	0·8310	0·8320	0·8329	0·8339	0·8348	0·8358	0·8368	0·8377	2	3	5	6	8
57	0·8387	0·8396	0·8406	0·8415	0·8425	0·8434	0·8443	0·8453	0·8462	0·8471	2	3	5	6	8
58	0·8480	0·8490	0·8499	0·8508	0·8517	0·8526	0·8536	0·8545	0·8554	0·8563	2	3	5	6	8
59	0·8572	0·8581	0·8590	0·8599	0·8607	0·8616	0·8625	0·8634	0·8643	0·8652	1	3	4	6	7
60	0·8660	0·8669	0·8678	0·8686	0·8695	0·8704	0·8712	0·8721	0·8729	0·8738	1	3	4	6	7
61	0·8746	0·8755	0·8763	0·8771	0·8780	0·8788	0·8796	0·8805	0·8813	0·8821	1	3	4	6	7
62	0·8829	0·8838	0·8846	0·8854	0·8862	0·8870	0·8878	0·8886	0·8894	0·8902	1	3	4	5	7
63	0·8910	0·8918	0·8926	0·8934	0·8942	0·8949	0·8957	0·8965	0·8973	0·8980	1	3	4	5	6
64	0·8988	0·8996	0·9003	0·9011	0·9018	0·9026	0·9033	0·9041	0·9048	0·9056	1	3	4	5	6
65	0·9063	0·9070	0·9078	0·9085	0·9092	0·9100	0·9107	0·9114	0·9121	0·9128	1	2	4	5	6
66	0·9135	0·9143	0·9150	0·9157	0·9164	0·9171	0·9178	0·9184	0·9191	0·9198	1	2	3	5	6
67	0·9205	0·9212	0·9219	0·9225	0·9232	0·9239	0·9245	0·9252	0·9259	0·9265	1	2	3	4	6
68	0·9272	0·9278	0·9285	0·9291	0·9298	0·9304	0·9311	0·9317	0·9323	0·9330	1	2	3	4	5
69	0·9336	0·9342	0·9348	0·9354	0·9361	0·9367	0·9373	0·9379	0·9385	0·9391	1	2	3	4	5
70	0·9397	0·9403	0·9409	0·9415	0·9421	0·9426	0·9432	0·9438	0·9444	0·9449	1	2	3	4	5
71	0·9455	0·9461	0·9466	0·9472	0·9478	0·9483	0·9489	0·9494	0·9500	0·9505	1	2	3	4	5
72	0·9511	0·9516	0·9521	0·9527	0·9532	0·9537	0·9542	0·9548	0·9553	0·9558	1	2	3	4	4
73	0·9563	0·9568	0·9573	0·9578	0·9583	0·9588	0·9593	0·9598	0·9603	0·9608	1	2	2	3	4
74	0·9613	0·9617	0·9622	0·9627	0·9632	0·9636	0·9641	0·9646	0·9650	0·9655	1	2	2	3	4
75	0·9659	0·9664	0·9668	0·9673	0·9677	0·9681	0·9686	0·9690	0·9694	0·9699	1	1	2	3	4
76	0·9703	0·9707	0·9711	0·9715	0·9720	0·9724	0·9728	0·9732	0·9736	0·9740	1	1	2	3	3
77	0·9744	0·9748	0·9751	0·9755	0·9759	0·9763	0·9767	0·9770	0·9774	0·9778	1	1	2	3	3
78	0·9781	0·9785	0·9789	0·9792	0·9796	0·9799	0·9803	0·9806	0·9810	0·9813	1	1	2	2	3
79	0·9816	0·9820	0·9823	0·9826	0·9829	0·9833	0·9836	0·9839	0·9842	0·9845	1	1	2	2	3
80	0·9848	0·9851	0·9854	0·9857	0·9860	0·9863	0·9866	0·9869	0·9871	0·9874	0	1	1	2	2
81	0·9877	0·9880	0·9882	0·9885	0·9888	0·9890	0·9893	0·9895	0·9898	0·9900	0	1	1	2	2
82	0·9903	0·9905	0·9907	0·9910	0·9912	0·9914	0·9917	0·9919	0·9921	0·9923	0	1	1	2	2
83	0·9925	0·9928	0·9930	0·9932	0·9934	0·9936	0·9938	0·9940	0·9942	0·9943	0	1	1	1	2
84	0·9945	0·9947	0·9949	0·9951	0·9952	0·9954	0·9956	0·9957	0·9959	0·9960	0	1	1	1	1
85	0·9962	0·9963	0·9965	0·9966	0·9968	0·9969	0·9971	0·9972	0·9973	0·9974	0	0	1	1	1
86	0·9976	0·9977	0·9978	0·9979	0·9980	0·9981	0·9982	0·9983	0·9984	0·9985	0	0	1	1	1
87	0·9986	0·9987	0·9988	0·9989	0·9990	0·9990	0·9991	0·9992	0·9993	0·9993	0	0	0	1	1
88	0·9994	0·9995	0·9995	0·9996	0·9996	0·9997	0·9997	0·9997	0·9998	0·9998	0	0	0	0	0
89	0·9998	0·9999	0·9999	0·9999	0·9999	1·0000	1·0000	1·0000	1·0000	1·0000	0	0	0	0	0

Table 2. Natural Cosines $\cos x^\circ$

x°	0°·0	0°·1	0°·2	0°·3	0°·4	0°·5	0°·6	0°·7	0°·8	0°·9	SUBTRACT				
	0′	6′	12′	18′	24′	30′	36′	42′	48′	54′	1′	2′	3′	4′	5′
0°	1·0000	1·0000	1·0000	1·0000	1·0000	1·0000	0·9999	0·9999	0·9999	0·9999	0	0	0	0	0
1	0·9998	0·9998	0·9998	0·9997	0·9997	0·9997	0·9996	0·9996	0·9995	0·9995	0	0	0	0	0
2	0·9994	0·9993	0·9993	0·9992	0·9991	0·9990	0·9990	0·9989	0·9988	0·9987	0	0	0	1	1
3	0·9986	0·9985	0·9984	0·9983	0·9982	0·9981	0·9980	0·9979	0·9978	0·9977	0	0	1	1	1
4	0·9976	0·9974	0·9973	0·9972	0·9971	0·9969	0·9968	0·9966	0·9965	0·9963	0	0	1	1	1
5	0·9962	0·9960	0·9959	0·9957	0·9956	0·9954	0·9952	0·9951	0·9949	0·9947	0	1	1	1	1
6	0·9945	0·9943	0·9942	0·9940	0·9938	0·9936	0·9934	0·9932	0·9930	0·9928	0	1	1	1	2
7	0·9925	0·9923	0·9921	0·9919	0·9917	0·9914	0·9912	0·9910	0·9907	0·9905	0	1	1	2	2
8	0·9903	0·9900	0·9898	0·9895	0·9893	0·9890	0·9888	0·9885	0·9882	0·9880	0	1	1	2	2
9	0·9877	0·9874	0·9871	0·9869	0·9866	0·9863	0·9860	0·9857	0·9854	0·9851	0	1	1	2	2
10	0·9848	0·9845	0·9842	0·9839	0·9836	0·9833	0·9829	0·9826	0·9823	0·9820	1	1	2	2	3
11	0·9816	0·9813	0·9810	0·9806	0·9803	0·9799	0·9796	0·9792	0·9789	0·9785	1	1	2	2	3
12	0·9781	0·9778	0·9774	0·9770	0·9767	0·9763	0·9759	0·9755	0·9751	0·9748	1	1	2	3	3
13	0·9744	0·9740	0·9736	0·9732	0·9728	0·9724	0·9720	0·9715	0·9711	0·9707	1	1	2	3	3
14	0·9703	0·9699	0·9694	0·9690	0·9686	0·9681	0·9677	0·9673	0·9668	0·9664	1	1	2	3	4
15	0·9659	0·9655	0·9650	0·9646	0·9641	0·9636	0·9632	0·9627	0·9622	0·9617	1	2	2	3	4
16	0·9613	0·9608	0·9603	0·9598	0·9593	0·9588	0·9583	0·9578	0·9573	0·9568	1	2	2	3	4
17	0·9563	0·9558	0·9553	0·9548	0·9542	0·9537	0·9532	0·9527	0·9521	0·9516	1	2	3	3	4
18	0·9511	0·9505	0·9500	0·9494	0·9489	0·9483	0·9478	0·9472	0·9466	0·9461	1	2	3	4	5
19	0·9455	0·9449	0·9444	0·9438	0·9432	0·9426	0·9421	0·9415	0·9409	0·9403	1	2	3	4	5
20	0·9397	0·9391	0·9385	0·9379	0·9373	0·9367	0·9361	0·9354	0·9348	0·9342	1	2	3	4	5
21	0·9336	0·9330	0·9323	0·9317	0·9311	0·9304	0·9298	0·9291	0·9285	0·9278	1	2	3	4	5
22	0·9272	0·9265	0·9259	0·9252	0·9245	0·9239	0·9232	0·9225	0·9219	0·9212	1	2	3	4	6
23	0·9205	0·9198	0·9191	0·9184	0·9178	0·9171	0·9164	0·9157	0·9150	0·9143	1	2	3	5	6
24	0·9135	0·9128	0·9121	0·9114	0·9107	0·9100	0·9092	0·9085	0·9078	0·9070	1	2	4	5	6
25	0·9063	0·9056	0·9048	0·9041	0·9033	0·9026	0·9018	0·9011	0·9003	0·8996	1	3	4	5	6
26	0·8988	0·8980	0·8973	0·8965	0·8957	0·8949	0·8942	0·8934	0·8926	0·8918	1	3	4	5	6
27	0·8910	0·8902	0·8894	0·8886	0·8878	0·8870	0·8862	0·8854	0·8846	0·8838	1	3	4	5	7
28	0·8829	0·8821	0·8813	0·8805	0·8796	0·8788	0·8780	0·8771	0·8763	0·8755	1	3	4	6	7
29	0·8746	0·8738	0·8729	0·8721	0·8712	0·8704	0·8695	0·8686	0·8678	0·8669	1	3	4	6	7
30	0·8660	0·8652	0·8643	0·8634	0·8625	0·8616	0·8607	0·8599	0·8590	0·8581	1	3	4	6	7
31	0·8572	0·8563	0·8554	0·8545	0·8536	0·8526	0·8517	0·8508	0·8499	0·8490	2	3	5	6	8
32	0·8480	0·8471	0·8462	0·8453	0·8443	0·8434	0·8425	0·8415	0·8406	0·8396	2	3	5	6	8
33	0·8387	0·8377	0·8368	0·8358	0·8348	0·8339	0·8329	0·8320	0·8310	0·8300	2	3	5	6	8
34	0·8290	0·8281	0·8271	0·8261	0·8251	0·8241	0·8231	0·8221	0·8211	0·8202	2	3	5	7	8
35	0·8192	0·8181	0·8171	0·8161	0·8151	0·8141	0·8131	0·8121	0·8111	0·8100	2	3	5	7	8
36	0·8090	0·8080	0·8070	0·8059	0·8049	0·8039	0·8028	0·8018	0·8007	0·7997	2	3	5	7	9
37	0·7986	0·7976	0·7965	0·7955	0·7944	0·7934	0·7923	0·7912	0·7902	0·7891	2	4	5	7	9
38	0·7880	0·7869	0·7859	0·7848	0·7837	0·7826	0·7815	0·7804	0·7793	0·7782	2	4	5	7	9
39	0·7771	0·7760	0·7749	0·7738	0·7727	0·7716	0·7705	0·7694	0·7683	0·7672	2	4	6	7	9
40	0·7660	0·7649	0·7638	0·7627	0·7615	0·7604	0·7593	0·7581	0·7570	0·7559	2	4	6	8	9
41	0·7547	0·7536	0·7524	0·7513	0·7501	0·7490	0·7478	0·7466	0·7455	0·7443	2	4	6	8	10
42	0·7431	0·7420	0·7408	0·7396	0·7385	0·7373	0·7361	0·7349	0·7337	0·7325	2	4	6	8	10
43	0·7314	0·7302	0·7290	0·7278	0·7266	0·7254	0·7242	0·7230	0·7218	0·7206	2	4	6	8	10
44	0·7193	0·7181	0·7169	0·7157	0·7145	0·7133	0·7120	0·7108	0·7096	0·7083	2	4	6	8	10

Natural Cosines $\cos x^\circ$

x°	0°·0 0′	0°·1 6′	0°·2 12′	0°·3 18′	0°·4 24′	0°·5 30′	0°·6 36′	0°·7 42′	0°·8 48′	0°·9 54′	SUBTRACT 1′	 2′	 3′	 4′	 5′
45°	0·7071	0·7059	0·7046	0·7034	0·7022	0·7009	0·6997	0·6984	0·6972	0·6959	2	4	6	8	10
46	0·6947	0·6934	0·6921	0·6909	0·6896	0·6884	0·6871	0·6858	0·6845	0·6833	2	4	6	8	11
47	0·6820	0·6807	0·6794	0·6782	0·6769	0·6756	0·6743	0·6730	0·6717	0·6704	2	4	6	9	11
48	0·6691	0·6678	0·6665	0·6652	0·6639	0·6626	0·6613	0·6600	0·6587	0·6574	2	4	7	9	11
49	0·6561	0·6547	0·6534	0·6521	0·6508	0·6494	0·6481	0·6468	0·6455	0·6441	2	4	7	9	11
50	0·6428	0·6414	0·6401	0·6388	0·6374	0·6361	0·6347	0·6334	0·6320	0·6307	2	4	7	9	11
51	0·6293	0·6280	0·6266	0·6252	0·6239	0·6225	0·6211	0·6198	0·6184	0·6170	2	5	7	9	11
52	0·6157	0·6143	0·6129	0·6115	0·6101	0·6088	0·6074	0·6060	0·6046	0·6032	2	5	7	9	12
53	0·6018	0·6004	0·5990	0·5976	0·5962	0·5948	0·5934	0·5920	0·5906	0·5892	2	5	7	9	12
54	0·5878	0·5864	0·5850	0·5835	0·5821	0·5807	0·5793	0·5779	0·5764	0·5750	2	5	7	9	12
55	0·5736	0·5721	0·5707	0·5693	0·5678	0·5664	0·5650	0·5635	0·5621	0·5606	2	5	7	10	12
56	0·5592	0·5577	0·5563	0·5548	0·5534	0·5519	0·5505	0·5490	0·5476	0·5461	2	5	7	10	12
57	0·5446	0·5432	0·5417	0·5402	0·5388	0·5373	0·5358	0·5344	0·5329	0·5314	2	5	7	10	12
58	0·5299	0·5284	0·5270	0·5255	0·5240	0·5225	0·5210	0·5195	0·5180	0·5165	2	5	7	10	12
59	0·5150	0·5135	0·5120	0·5105	0·5090	0·5075	0·5060	0·5045	0·5030	0·5015	3	5	8	10	13
60	0·5000	0·4985	0·4970	0·4955	0·4939	0·4924	0·4909	0·4894	0·4879	0·4863	3	5	8	10	13
61	0·4848	0·4833	0·4818	0·4802	0·4787	0·4772	0·4756	0·4741	0·4726	0·4710	3	5	8	10	13
62	0·4695	0·4679	0·4664	0·4648	0·4633	0·4617	0·4602	0·4586	0·4571	0·4555	3	5	8	10	13
63	0·4540	0·4524	0·4509	0·4493	0·4478	0·4462	0·4446	0·4431	0·4415	0·4399	3	5	8	10	13
64	0·4384	0·4368	0·4352	0·4337	0·4321	0·4305	0·4289	0·4274	0·4258	0·4242	3	5	8	11	13
65	0·4226	0·4210	0·4195	0·4179	0·4163	0·4147	0·4131	0·4115	0·4099	0·4083	3	5	8	11	13
66	0·4067	0·4051	0·4035	0·4019	0·4003	0·3987	0·3971	0·3955	0·3939	0·3923	3	5	8	11	13
67	0·3907	0·3891	0·3875	0·3859	0·3843	0·3827	0·3811	0·3795	0·3778	0·3762	3	5	8	11	13
68	0·3746	0·3730	0·3714	0·3697	0·3681	0·3665	0·3649	0·3633	0·3616	0·3600	3	5	8	11	14
69	0·3584	0·3567	0·3551	0·3535	0·3518	0·3502	0·3486	0·3469	0·3453	0·3437	3	5	8	11	14
70	0·3420	0·3404	0·3387	0·3371	0·3355	0·3338	0·3322	0·3305	0·3289	0·3272	3	5	8	11	14
71	0·3256	0·3239	0·3223	0·3206	0·3190	0·3173	0·3156	0·3140	0·3123	0·3107	3	6	8	11	14
72	0·3090	0·3074	0·3057	0·3040	0·3024	0·3007	0·2990	0·2974	0·2957	0·2940	3	6	8	11	14
73	0·2924	0·2907	0·2890	0·2874	0·2857	0·2840	0·2823	0·2807	0·2790	0·2773	3	6	8	11	14
74	0·2756	0·2740	0·2723	0·2706	0·2689	0·2672	0·2656	0·2639	0·2622	0·2605	3	6	8	11	14
75	0·2588	0·2571	0·2554	0·2538	0·2521	0·2504	0·2487	0·2470	0·2453	0·2436	3	6	8	11	14
76	0·2419	0·2402	0·2385	0·2368	0·2351	0·2334	0·2317	0·2300	0·2284	0·2267	3	6	8	11	14
77	0·2250	0·2233	0·2215	0·2198	0·2181	0·2164	0·2147	0·2130	0·2113	0·2096	3	6	9	11	14
78	0·2079	0·2062	0·2045	0·2028	0·2011	0·1994	0·1977	0·1959	0·1942	0·1925	3	6	9	11	14
79	0·1908	0·1891	0·1874	0·1857	0·1840	0·1822	0·1805	0·1788	0·1771	0·1754	3	6	9	11	14
80	0·1736	0·1719	0·1702	0·1685	0·1668	0·1650	0·1633	0·1616	0·1599	0·1582	3	6	9	11	14
81	0·1564	0·1547	0·1530	0·1513	0·1495	0·1478	0·1461	0·1444	0·1426	0·1409	3	6	9	12	14
82	0·1392	0·1374	0·1357	0·1340	0·1323	0·1305	0·1288	0·1271	0·1253	0·1236	3	6	9	12	14
83	0·1219	0·1201	0·1184	0·1167	0·1149	0·1132	0·1115	0·1097	0·1080	0·1063	3	6	9	12	14
84	0·1045	0·1028	0·1011	0·0993	0·0976	0·0958	0·0941	0·0924	0·0906	0·0889	3	6	9	12	14
85	0·0872	0·0854	0·0837	0·0819	0·0802	0·0785	0·0767	0·0750	0·0732	0·0715	3	6	9	12	14
86	0·0698	0·0680	0·0663	0·0645	0·0628	0·0610	0·0593	0·0576	0·0558	0·0541	3	6	9	12	15
87	0·0523	0·0506	0·0488	0·0471	0·0454	0·0436	0·0419	0·0401	0·0384	0·0366	3	6	9	12	15
88	0·0349	0·0332	0·0314	0·0297	0·0279	0·0262	0·0244	0·0227	0·0209	0·0192	3	6	9	12	15
89	0·0175	0·0157	0·0140	0·0122	0·0105	0·0087	0·0070	0·0052	0·0035	0·0017	3	6	9	12	15

Table 3. Natural Tangents $\tan x^\circ$

x°	0°·0 0′	0°·1 6′	0°·2 12′	0°·3 18′	0°·4 24′	0°·5 30′	0°·6 36′	0°·7 42′	0°·8 48′	0°·9 54′	1′	2′	3′	4′	5′
0°	0·0000	0·0017	0·0035	0·0052	0·0070	0·0087	0·0105	0·0122	0·0140	0·0157	3	6	9	12	15
1	0·0175	0·0192	0·0209	0·0227	0·0244	0·0262	0·0279	0·0297	0·0314	0·0332	3	6	9	12	15
2	0·0349	0·0367	0·0384	0·0402	0·0419	0·0437	0·0454	0·0472	0·0489	0·0507	3	6	9	12	15
3	0·0524	0·0542	0·0559	0·0577	0·0594	0·0612	0·0629	0·0647	0·0664	0·0682	3	6	9	12	15
4	0·0699	0·0717	0·0734	0·0752	0·0769	0·0787	0·0805	0·0822	0·0840	0·0857	3	6	9	12	15
5	0·0875	0·0892	0·0910	0·0928	0·0945	0·0963	0·0981	0·0998	0·1016	0·1033	3	6	9	12	15
6	0·1051	0·1069	0·1086	0·1104	0·1122	0·1139	0·1157	0·1175	0·1192	0·1210	3	6	9	12	15
7	0·1228	0·1246	0·1263	0·1281	0·1299	0·1317	0·1334	0·1352	0·1370	0·1388	3	6	9	12	15
8	0·1405	0·1423	0·1441	0·1459	0·1477	0·1495	0·1512	0·1530	0·1548	0·1566	3	6	9	12	15
9	0·1584	0·1602	0·1620	0·1638	0·1655	0·1673	0·1691	0·1709	0·1727	0·1745	3	6	9	12	15
10	0·1763	0·1781	0·1799	0·1817	0·1835	0·1853	0·1871	0·1890	0·1908	0·1926	3	6	9	12	15
11	0·1944	0·1962	0·1980	0·1998	0·2016	0·2035	0·2053	0·2071	0·2089	0·2107	3	6	9	12	15
12	0·2126	0·2144	0·2162	0·2180	0·2199	0·2217	0·2235	0·2254	0·2272	0·2290	3	6	9	12	15
13	0·2309	0·2327	0·2345	0·2364	0·2382	0·2401	0·2419	0·2438	0·2456	0·2475	3	6	9	12	15
14	0·2493	0·2512	0·2530	0·2549	0·2568	0·2586	0·2605	0·2623	0·2642	0·2661	3	6	9	12	16
15	0·2679	0·2698	0·2717	0·2736	0·2754	0·2773	0·2792	0·2811	0·2830	0·2849	3	6	9	13	16
16	0·2867	0·2886	0·2905	0·2924	0·2943	0·2962	0·2981	0·3000	0·3019	0·3038	3	6	9	13	16
17	0·3057	0·3076	0·3096	0·3115	0·3134	0·3153	0·3172	0·3191	0·3211	0·3230	3	6	10	13	16
18	0·3249	0·3269	0·3288	0·3307	0·3327	0·3346	0·3365	0·3385	0·3404	0·3424	3	6	10	13	16
19	0·3443	0·3463	0·3482	0·3502	0·3522	0·3541	0·3561	0·3581	0·3600	0·3620	3	7	10	13	16
20	0·3640	0·3659	0·3679	0·3699	0·3719	0·3739	0·3759	0·3779	0·3799	0·3819	3	7	10	13	17
21	0·3839	0·3859	0·3879	0·3899	0·3919	0·3939	0·3959	0·3979	0·4000	0·4020	3	7	10	13	17
22	0·4040	0·4061	0·4081	0·4101	0·4122	0·4142	0·4163	0·4183	0·4204	0·4224	3	7	10	14	17
23	0·4245	0·4265	0·4286	0·4307	0·4327	0·4348	0·4369	0·4390	0·4411	0·4431	3	7	10	14	17
24	0·4452	0·4473	0·4494	0·4515	0·4536	0·4557	0·4578	0·4599	0·4621	0·4642	4	7	11	14	18
25	0·4663	0·4684	0·4706	0·4727	0·4748	0·4770	0·4791	0·4813	0·4834	0·4856	4	7	11	14	18
26	0·4877	0·4899	0·4921	0·4942	0·4964	0·4986	0·5008	0·5029	0·5051	0·5073	4	7	11	15	18
27	0·5095	0·5117	0·5139	0·5161	0·5184	0·5206	0·5228	0·5250	0·5272	0·5295	4	7	11	15	18
28	0·5317	0·5340	0·5362	0·5384	0·5407	0·5430	0·5452	0·5475	0·5498	0·5520	4	8	11	15	19
29	0·5543	0·5566	0·5589	0·5612	0·5635	0·5658	0·5681	0·5704	0·5727	0·5750	4	8	12	15	19
30	0·5774	0·5797	0·5820	0·5844	0·5867	0·5890	0·5914	0·5938	0·5961	0·5985	4	8	12	16	20
31	0·6009	0·6032	0·6056	0·6080	0·6104	0·6128	0·6152	0·6176	0·6200	0·6224	4	8	12	16	20
32	0·6249	0·6273	0·6297	0·6322	0·6346	0·6371	0·6395	0·6420	0·6445	0·6469	4	8	12	16	20
33	0·6494	0·6519	0·6544	0·6569	0·6594	0·6619	0·6644	0·6669	0·6694	0·6720	4	8	13	17	21
34	0·6745	0·6771	0·6796	0·6822	0·6847	0·6873	0·6899	0·6924	0·6950	0·6976	4	9	13	17	21
35	0·7002	0·7028	0·7054	0·7080	0·7107	0·7133	0·7159	0·7186	0·7212	0·7239	4	9	13	18	22
36	0·7265	0·7292	0·7319	0·7346	0·7373	0·7400	0·7427	0·7454	0·7481	0·7508	4	9	13	18	23
37	0·7536	0·7563	0·7590	0·7618	0·7646	0·7673	0·7701	0·7729	0·7757	0·7785	5	9	14	18	23
38	0·7813	0·7841	0·7869	0·7898	0·7926	0·7954	0·7983	0·8012	0·8040	0·8069	5	9	14	19	24
39	0·8098	0·8127	0·8156	0·8185	0·8214	0·8243	0·8273	0·8302	0·8332	0·8361	5	10	15	20	24
40	0·8391	0·8421	0·8451	0·8481	0·8511	0·8541	0·8571	0·8601	0·8632	0·8662	5	10	15	20	25
41	0·8693	0·8724	0·8754	0·8785	0·8816	0·8847	0·8878	0·8910	0·8941	0·8972	5	10	16	21	26
42	0·9004	0·9036	0·9067	0·9099	0·9131	0·9163	0·9195	0·9228	0·9260	0·9293	5	11	16	21	27
43	0·9325	0·9358	0·9391	0·9424	0·9457	0·9490	0·9523	0·9556	0·9590	0·9623	6	11	17	22	28
44	0·9657	0·9691	0·9725	0·9759	0·9793	0·9827	0·9861	0·9896	0·9930	0·9965	6	11	17	23	29

Natural Tangents tan x°

x°	0°·0 0′	0°·1 6′	0°·2 12′	0°·3 18′	0°·4 24′	0°·5 30′	0°·6 36′	0°·7 42′	0°·8 48′	0°·9 54′	1′	2′	3′	4′	5′
45	1·0000	1·0035	1·0070	1·0105	1·0141	1·0176	1·0212	1·0247	1·0283	1·0319	6	12	18	24	30
46	1·0355	1·0392	1·0428	1·0464	1·0501	1·0538	1·0575	1·0612	1·0649	1·0686	6	12	18	25	31
47	1·0724	1·0761	1·0799	1·0837	1·0875	1·0913	1·0951	1·0990	1·1028	1·1067	6	13	19	25	32
48	1·1106	1·1145	1·1184	1·1224	1·1263	1·1303	1·1343	1·1383	1·1423	1·1463	7	13	20	26	33
49	1·1504	1·1544	1·1585	1·1626	1·1667	1·1708	1·1750	1·1792	1·1833	1·1875	7	14	21	28	34
50	1·1918	1·1960	1·2002	1·2045	1·2088	1·2131	1·2174	1·2218	1·2261	1·2305	7	14	22	29	36
51	1·2349	1·2393	1·2437	1·2482	1·2527	1·2572	1·2617	1·2662	1·2708	1·2753	7	15	22	30	38
52	1·2799	1·2846	1·2892	1·2938	1·2985	1·3032	1·3079	1·3127	1·3175	1·3222	8	16	24	31	39
53	1·3270	1·3319	1·3367	1·3416	1·3465	1·3514	1·3564	1·3613	1·3663	1·3713	8	16	25	33	41
54	1·3764	1·3814	1·3865	1·3916	1·3968	1·4019	1·4071	1·4124	1·4176	1·4229	9	17	26	34	43
55	1·4281	1·4335	1·4388	1·4442	1·4496	1·4550	1·4605	1·4659	1·4715	1·4770	9	18	27	36	45
56	1·4826	1·4882	1·4938	1·4994	1·5051	1·5108	1·5166	1·5224	1·5282	1·5340	10	19	29	38	48
57	1·5399	1·5458	1·5517	1·5577	1·5637	1·5697	1·5757	1·5818	1·5880	1·5941	10	20	30	40	50
58	1·6003	1·6066	1·6128	1·6191	1·6255	1·6319	1·6383	1·6447	1·6512	1·6577	11	21	32	43	53
59	1·6643	1·6709	1·6775	1·6842	1·6909	1·6977	1·7045	1·7113	1·7182	1·7251	11	23	34	45	56
60	1·7321	1·7391	1·7461	1·7532	1·7603	1·7675	1·7747	1·7820	1·7893	1·7966	12	24	36	48	60
61	1·8040	1·8115	1·8190	1·8265	1·8341	1·8418	1·8495	1·8572	1·8650	1·8728	13	26	38	51	64
62	1·8807	1·8887	1·8967	1·9047	1·9128	1·9210	1·9292	1·9375	1·9458	1·9542	14	27	41	55	68
63	1·9626	1·9711	1·9797	1·9883	1·9970	2·0057	2·0145	2·0233	2·0323	2·0413	15	29	44	58	73
64	2·0503	2·0594	2·0686	2·0778	2·0872	2·0965	2·1060	2·1155	2·1251	2·1348	16	31	47	63	78
65	2·1445	2·1543	2·1642	2·1742	2·1842	2·1943	2·2045	2·2148	2·2251	2·2355	17	34	51	68	85
66	2·2460	2·2566	2·2673	2·2781	2·2889	2·2998	2·3109	2·3220	2·3332	2·3445	18	37	55	73	91
67	2·3559	2·3673	2·3789	2·3906	2·4023	2·4142	2·4262	2·4383	2·4504	2·4627	20	40	59	79	99
68	2·4751	2·4876	2·5002	2·5129	2·5257	2·5386	2·5517	2·5649	2·5782	2·5916	22	43	65	87	108
69	2·6051	2·6187	2·6325	2·6464	2·6605	2·6746	2·6889	2·7034	2·7179	2·7326	24	47	71	95	119
70	2·7475	2·7625	2·7776	2·7929	2·8083	2·8239	2·8397	2·8556	2·8716	2·8878	26	52	78	104	131
71	2·9042	2·9208	2·9375	2·9544	2·9714	2·9887	3·0061	3·0237	3·0415	3·0595	29	58	87	115	144
72	3·0777	3·0961	3·1146	3·1334	3·1524	3·1716	3·1910	3·2106	3·2305	3·2506	32	64	96	129	161
73	3·2709	3·2914	3·3122	3·3332	3·3544	3·3759	3·3977	3·4197	3·4420	3·4646	36	72	108	144	180
74	3·4874	3·5105	3·5339	3·5576	3·5816	3·6059	3·6305	3·6554	3·6806	3·7062	41	81	122	163	204
75	3·7321	3·7583	3·7848	3·8118	3·8391	3·8667	3·8947	3·9232	3·9520	3·9812	46	92	139	185	232
76	4·0108	4·0408	4·0713	4·1022	4·1335	4·1653	4·1976	4·2303	4·2635	4·2972	53	106	160	213	267
77	4·3315	4·3662	4·4015	4·4373	4·4737	4·5107	4·5483	4·5864	4·6252	4·6646	62	124	186	248	311
78	4·7046	4·7453	4·7867	4·8288	4·8716	4·9152	4·9594	5·0045	5·0504	5·0970	73	146	219	292	366
79	5·1446	5·1929	5·2422	5·2924	5·3435	5·3955	5·4486	5·5026	5·5578	5·6140	87	174	262	350	438
80	5·6713	5·7297	5·7894	5·8502	5·9124	5·9758	6·0405	6·1066	6·1742	6·2432					
81	6·3138	6·3859	6·4596	6·5350	6·6122	6·6912	6·7720	6·8548	6·9395	7·0264					
82	7·1154	7·2066	7·3002	7·3962	7·4947	7·5958	7·6996	7·8062	7·9158	8·0285					
83	8·1443	8·2636	8·3863	8·5126	8·6427	8·7769	8·9152	9·0579	9·2052	9·3572					
84	9·5144	9·6768	9·8448	10·019	10·199	10·385	10·579	10·780	10·988	11·205					
85	11·430	11·665	11·909	12·163	12·429	12·706	12·996	13·300	13·617	13·951					
86	14·301	14·669	15·056	15·464	15·895	16·350	16·832	17·343	17·886	18·465					
87	19·081	19·740	20·447	21·205	22·022	22·904	23·859	24·898	26·031	27·272					
88	28·636	30·145	31·821	33·694	35·801	38·189	40·917	44·066	47·740	52·081					
89	57·290	63·657	71·615	81·847	95·490	114·59	143·24	190·98	286·48	572·96					

*

Table 4. Logarithms $\log_{10} x$

x	0	1	2	3	4	5	6	7	8	9	1	2	3	4	5	6	7	8	9
1·0	0·0000	0·0043	0·0086	0·0128	0·0170	0·0212	0·0253	0·0294	0·0334	0·0374	4	8	13	17	21	25	29	33	37
1·1	0·0414	0·0453	0·0492	0·0531	0·0569	0·0607	0·0645	0·0682	0·0719	0·0755	4	8	11	15	19	23	27	30	34
1·2	0·0792	0·0828	0·0864	0·0899	0·0934	0·0969	0·1004	0·1038	0·1072	0·1106	4	7	10	14	17	21	24	28	31
1·3	0·1139	0·1173	0·1206	0·1239	0·1271	0·1303	0·1335	0·1367	0·1399	0·1430	3	6	10	13	16	19	23	26	29
1·4	0·1461	0·1492	0·1523	0·1553	0·1584	0·1614	0·1644	0·1673	0·1703	0·1732	3	6	9	12	15	18	21	24	27
1·5	0·1761	0·1790	0·1818	0·1847	0·1875	0·1903	0·1931	0·1959	0·1987	0·2014	3	6	8	11	14	17	20	23	25
1·6	0·2041	0·2068	0·2095	0·2122	0·2148	0·2175	0·2201	0·2227	0·2253	0·2279	3	5	8	11	13	16	18	21	24
1·7	0·2304	0·2330	0·2355	0·2380	0·2405	0·2430	0·2455	0·2480	0·2504	0·2529	2	5	7	10	12	15	17	20	22
1·8	0·2553	0·2577	0·2601	0·2625	0·2648	0·2672	0·2695	0·2718	0·2742	0·2765	2	5	7	9	12	14	16	19	21
1·9	0·2788	0·2810	0·2833	0·2856	0·2878	0·2900	0·2923	0·2945	0·2967	0·2989	2	4	7	9	11	13	16	18	20
2·0	0·3010	0·3032	0·3054	0·3075	0·3096	0·3118	0·3139	0·3160	0·3181	0·3201	2	4	6	9	11	13	15	17	19
2·1	0·3222	0·3243	0·3263	0·3284	0·3304	0·3324	0·3345	0·3365	0·3385	0·3404	2	4	6	8	10	12	14	16	18
2·2	0·3424	0·3444	0·3464	0·3483	0·3502	0·3522	0·3541	0·3560	0·3579	0·3598	2	4	6	8	10	12	14	15	17
2·3	0·3617	0·3636	0·3655	0·3674	0·3692	0·3711	0·3729	0·3747	0·3766	0·3784	2	4	6	7	9	11	13	15	17
2·4	0·3802	0·3820	0·3838	0·3856	0·3874	0·3892	0·3909	0·3927	0·3945	0·3962	2	4	5	7	9	11	12	14	16
2·5	0·3979	0·3997	0·4014	0·4031	0·4048	0·4065	0·4082	0·4099	0·4116	0·4133	2	3	5	7	9	10	12	14	15
2·6	0·4150	0·4166	0·4183	0·4200	0·4216	0·4232	0·4249	0·4265	0·4281	0·4298	2	3	5	7	8	10	12	13	15
2·7	0·4314	0·4330	0·4346	0·4362	0·4378	0·4393	0·4409	0·4425	0·4440	0·4456	2	3	5	6	8	9	11	13	14
2·8	0·4472	0·4487	0·4502	0·4518	0·4533	0·4548	0·4564	0·4579	0·4594	0·4609	2	3	5	6	8	9	11	12	14
2·9	0·4624	0·4639	0·4654	0·4669	0·4683	0·4698	0·4713	0·4728	0·4742	0·4757	1	3	4	6	7	9	10	12	13
3·0	0·4771	0·4786	0·4800	0·4814	0·4829	0·4843	0·4857	0·4871	0·4886	0·4900	1	3	4	6	7	9	10	11	13
3·1	0·4914	0·4928	0·4942	0·4955	0·4969	0·4983	0·4997	0·5011	0·5024	0·5038	1	3	4	6	7	8	10	11	12
3·2	0·5052	0·5065	0·5079	0·5092	0·5105	0·5119	0·5132	0·5145	0·5159	0·5172	1	3	4	5	7	8	9	11	12
3·3	0·5185	0·5198	0·5211	0·5224	0·5237	0·5250	0·5263	0·5276	0·5289	0·5302	1	3	4	5	6	8	9	10	12
3·4	0·5315	0·5328	0·5340	0·5353	0·5366	0·5378	0·5391	0·5403	0·5416	0·5428	1	3	4	5	6	8	9	10	11
3·5	0·5441	0·5453	0·5465	0·5478	0·5490	0·5502	0·5514	0·5527	0·5539	0·5551	1	2	4	5	6	7	9	10	11
3·6	0·5563	0·5575	0·5587	0·5599	0·5611	0·5623	0·5635	0·5647	0·5658	0·5670	1	2	4	5	6	7	8	10	11
3·7	0·5682	0·5694	0·5705	0·5717	0·5729	0·5740	0·5752	0·5763	0·5775	0·5786	1	2	3	5	6	7	8	9	10
3·8	0·5798	0·5809	0·5821	0·5832	0·5843	0·5855	0·5866	0·5877	0·5888	0·5899	1	2	3	5	6	7	8	9	10
3·9	0·5911	0·5922	0·5933	0·5944	0·5955	0·5966	0·5977	0·5988	0·5999	0·6010	1	2	3	4	6	7	8	9	10
4·0	0·6021	0·6031	0·6042	0·6053	0·6064	0·6075	0·6085	0·6096	0·6107	0·6117	1	2	3	4	5	6	8	9	10
4·1	0·6128	0·6138	0·6149	0·6159	0·6170	0·6180	0·6191	0·6201	0·6212	0·6222	1	2	3	4	5	6	7	8	9
4·2	0·6232	0·6243	0·6253	0·6263	0·6274	0·6284	0·6294	0·6304	0·6314	0·6325	1	2	3	4	5	6	7	8	9
4·3	0·6335	0·6345	0·6355	0·6365	0·6375	0·6385	0·6395	0·6405	0·6415	0·6425	1	2	3	4	5	6	7	8	9
4·4	0·6435	0·6444	0·6454	0·6464	0·6474	0·6484	0·6493	0·6503	0·6513	0·6522	1	2	3	4	5	6	7	8	9
4·5	0·6532	0·6542	0·6551	0·6561	0·6571	0·6580	0·6590	0·6599	0·6609	0·6618	1	2	3	4	5	6	7	8	9
4·6	0·6628	0·6637	0·6646	0·6656	0·6665	0·6675	0·6684	0·6693	0·6702	0·6712	1	2	3	4	5	6	7	7	8
4·7	0·6721	0·6730	0·6739	0·6749	0·6758	0·6767	0·6776	0·6785	0·6794	0·6803	1	2	3	4	5	5	6	7	8
4·8	0·6812	0·6821	0·6830	0·6839	0·6848	0·6857	0·6866	0·6875	0·6884	0·6893	1	2	3	4	4	5	6	7	8
4·9	0·6902	0·6911	0·6920	0·6928	0·6937	0·6946	0·6955	0·6964	0·6972	0·6981	1	2	3	4	4	5	6	7	8
5·0	0·6990	0·6998	0·7007	0·7016	0·7024	0·7033	0·7042	0·7050	0·7059	0·7067	1	2	3	3	4	5	6	7	8
5·1	0·7076	0·7084	0·7093	0·7101	0·7110	0·7118	0·7126	0·7135	0·7143	0·7152	1	2	3	3	4	5	6	7	8
5·2	0·7160	0·7168	0·7177	0·7185	0·7193	0·7202	0·7210	0·7218	0·7226	0·7235	1	2	2	3	4	5	6	7	7
5·3	0·7243	0·7251	0·7259	0·7267	0·7275	0·7284	0·7292	0·7300	0·7308	0·7316	1	2	2	3	4	5	6	7	7
5·4	0·7324	0·7332	0·7340	0·7348	0·7356	0·7364	0·7372	0·7380	0·7388	0·7396	1	2	2	3	4	5	6	6	7

x	0	1	2	3	4	5	6	7	8	9	1	2	3	4	5	6	7	8	9
5·5	0·7404	0·7412	0·7419	0·7427	0·7435	0·7443	0·7451	0·7459	0·7466	0·7474	1	2	2	3	4	5	5	6	7
5·6	0·7482	0·7490	0·7497	0·7505	0·7513	0·7520	0·7528	0·7536	0·7543	0·7551	1	2	2	3	4	5	5	6	7
5·7	0·7559	0·7566	0·7574	0·7582	0·7589	0·7597	0·7604	0·7612	0·7619	0·7627	1	2	2	3	4	5	5	6	7
5·8	0·7634	0·7642	0·7649	0·7657	0·7664	0·7672	0·7679	0·7686	0·7694	0·7701	1	1	2	3	4	4	5	6	7
5·9	0·7709	0·7716	0·7723	0·7731	0·7738	0·7745	0·7752	0·7760	0·7767	0·7774	1	1	2	3	4	4	5	6	7
6·0	0·7782	0·7789	0·7796	0·7803	0·7810	0·7818	0·7825	0·7832	0·7839	0·7846	1	1	2	3	4	4	5	6	6
6·1	0·7853	0·7860	0·7868	0·7875	0·7882	0·7889	0·7896	0·7903	0·7910	0·7917	1	1	2	3	4	4	5	6	6
6·2	0·7924	0·7931	0·7938	0·7945	0·7952	0·7959	0·7966	0·7973	0·7980	0·7987	1	1	2	3	3	4	5	6	6
6·3	0·7993	0·8000	0·8007	0·8014	0·8021	0·8028	0·8035	0·8041	0·8048	0·8055	1	1	2	3	3	4	5	5	6
6·4	0·8062	0·8069	0·8075	0·8082	0·8089	0·8096	0·8102	0·8109	0·8116	0·8122	1	1	2	3	3	4	5	5	6
6·5	0·8129	0·8136	0·8142	0·8149	0·8156	0·8162	0·8169	0·8176	0·8182	0·8189	1	1	2	3	3	4	5	5	6
6·6	0·8195	0·8202	0·8209	0·8215	0·8222	0·8228	0·8235	0·8241	0·8248	0·8254	1	1	2	3	3	4	5	5	6
6·7	0·8261	0·8267	0·8274	0·8280	0·8287	0·8293	0·8299	0·8306	0·8312	0·8319	1	1	2	3	3	4	5	5	6
6·8	0·8325	0·8331	0·8338	0·8344	0·8351	0·8357	0·8363	0·8370	0·8376	0·8382	1	1	2	3	3	4	4	5	6
6·9	0·8388	0·8395	0·8401	0·8407	0·8414	0·8420	0·8426	0·8432	0·8439	0·8445	1	1	2	3	3	4	4	5	6
7·0	0·8451	0·8457	0·8463	0·8470	0·8476	0·8482	0·8488	0·8494	0·8500	0·8506	1	1	2	2	3	4	4	5	6
7·1	0·8513	0·8519	0·8525	0·8531	0·8537	0·8543	0·8549	0·8555	0·8561	0·8567	1	1	2	2	3	4	4	5	5
7·2	0·8573	0·8579	0·8585	0·8591	0·8597	0·8603	0·8609	0·8615	0·8621	0·8627	1	1	2	2	3	4	4	5	5
7·3	0·8633	0·8639	0·8645	0·8651	0·8657	0·8663	0·8669	0·8675	0·8681	0·8686	1	1	2	2	3	4	4	5	5
7·4	0·8692	0·8698	0·8704	0·8710	0·8716	0·8722	0·8727	0·8733	0·8739	0·8745	1	1	2	2	3	4	4	5	5
7·5	0·8751	0·8756	0·8762	0·8768	0·8774	0·8779	0·8785	0·8791	0·8797	0·8802	1	1	2	2	3	3	4	5	5
7·6	0·8808	0·8814	0·8820	0·8825	0·8831	0·8837	0·8842	0·8848	0·8854	0·8859	1	1	2	2	3	3	4	5	5
7·7	0·8865	0·8871	0·8876	0·8882	0·8887	0·8893	0·8899	0·8904	0·8910	0·8915	1	1	2	2	3	3	4	4	5
7·8	0·8921	0·8927	0·8932	0·8938	0·8943	0·8949	0·8954	0·8960	0·8965	0·8971	1	1	2	2	3	3	4	4	5
7·9	0·8976	0·8982	0·8987	0·8993	0·8998	0·9004	0·9009	0·9015	0·9020	0·9025	1	1	2	2	3	3	4	4	5
8·0	0·9031	0·9036	0·9042	0·9047	0·9053	0·9058	0·9063	0·9069	0·9074	0·9079	1	1	2	2	3	3	4	4	5
8·1	0·9085	0·9090	0·9096	0·9101	0·9106	0·9112	0·9117	0·9122	0·9128	0·9133	1	1	2	2	3	3	4	4	5
8·2	0·9138	0·9143	0·9149	0·9154	0·9159	0·9165	0·9170	0·9175	0·9180	0·9186	1	1	2	2	3	3	4	4	5
8·3	0·9191	0·9196	0·9201	0·9206	0·9212	0·9217	0·9222	0·9227	0·9232	0·9238	1	1	2	2	3	3	4	4	5
8·4	0·9243	0·9248	0·9253	0·9258	0·9263	0·9269	0·9274	0·9279	0·9284	0·9289	1	1	2	2	3	3	4	4	5
8·5	0·9294	0·9299	0·9304	0·9309	0·9315	0·9320	0·9325	0·9330	0·9335	0·9340	1	1	2	2	3	3	4	4	5
8·6	0·9345	0·9350	0·9355	0·9360	0·9365	0·9370	0·9375	0·9380	0·9385	0·9390	1	1	2	2	3	3	4	4	5
8·7	0·9395	0·9400	0·9405	0·9410	0·9415	0·9420	0·9425	0·9430	0·9435	0·9440	0	1	1	2	2	3	3	4	4
8·8	0·9445	0·9450	0·9455	0·9460	0·9465	0·9469	0·9474	0·9479	0·9484	0·9489	0	1	1	2	2	3	3	4	4
8·9	0·9494	0·9499	0·9504	0·9509	0·9513	0·9518	0·9523	0·9528	0·9533	0·9538	0	1	1	2	2	3	3	4	4
9·0	0·9542	0·9547	0·9552	0·9557	0·9562	0·9566	0·9571	0·9576	0·9581	0·9586	0	1	1	2	2	3	3	4	4
9·1	0·9590	0·9595	0·9600	0·9605	0·9609	0·9614	0·9619	0·9624	0·9628	0·9633	0	1	1	2	2	3	3	4	4
9·2	0·9638	0·9643	0·9647	0·9652	0·9657	0·9661	0·9666	0·9671	0·9675	0·9680	0	1	1	2	2	3	3	4	4
9·3	0·9685	0·9689	0·9694	0·9699	0·9703	0·9708	0·9713	0·9717	0·9722	0·9727	0	1	1	2	2	3	3	4	4
9·4	0·9731	0·9736	0·9741	0·9745	0·9750	0·9754	0·9759	0·9763	0·9768	0·9773	0	1	1	2	2	3	3	4	4
9·5	0·9777	0·9782	0·9786	0·9791	0·9795	0·9800	0·9805	0·9809	0·9814	0·9818	0	1	1	2	2	3	3	4	4
9·6	0·9823	0·9827	0·9832	0·9836	0·9841	0·9845	0·9850	0·9854	0·9859	0·9863	0	1	1	2	2	3	3	4	4
9·7	0·9868	0·9872	0·9877	0·9881	0·9886	0·9890	0·9894	0·9899	0·9903	0·9908	0	1	1	2	2	3	3	4	4
9·8	0·9912	0·9917	0·9921	0·9926	0·9930	0·9934	0·9939	0·9943	0·9948	0·9952	0	1	1	2	2	3	3	4	4
9·9	0·9956	0·9961	0·9965	0·9969	0·9974	0·9978	0·9983	0·9987	0·9991	0·9996	0	1	1	2	2	3	3	3	4

Table 5. Logarithms of Sines $\log_{10} \sin x^\circ$

x°	0°·0 0′	0°·1 6′	0°·2 12′	0°·3 18′	0°·4 24′	0°·5 30′	0°·6 36′	0°·7 42′	0°·8 48′	0°·9 54′	1′	2′	3′	4′	5′
0°	−∞	$\bar{3}$·2419	$\bar{3}$·5429	$\bar{3}$·7190	$\bar{3}$·8439	$\bar{3}$·9408	$\bar{2}$·0200	$\bar{2}$·0870	$\bar{2}$·1450	$\bar{2}$·1961					
1	$\bar{2}$·2419	$\bar{2}$·2832	$\bar{2}$·3210	$\bar{2}$·3558	$\bar{2}$·3880	$\bar{2}$·4179	$\bar{2}$·4459	$\bar{2}$·4723	$\bar{2}$·4971	$\bar{2}$·5206					
2	$\bar{2}$·5428	$\bar{2}$·5640	$\bar{2}$·5842	$\bar{2}$·6035	$\bar{2}$·6220	$\bar{2}$·6397	$\bar{2}$·6567	$\bar{2}$·6731	$\bar{2}$·6889	$\bar{2}$·7041	30	59	89	118	147
3	$\bar{2}$·7188	$\bar{2}$·7330	$\bar{2}$·7468	$\bar{2}$·7602	$\bar{2}$·7731	$\bar{2}$·7857	$\bar{2}$·7979	$\bar{2}$·8098	$\bar{2}$·8213	$\bar{2}$·8326	21	42	63	84	104
4	$\bar{2}$·8436	$\bar{2}$·8543	$\bar{2}$·8647	$\bar{2}$·8749	$\bar{2}$·8849	$\bar{2}$·8946	$\bar{2}$·9042	$\bar{2}$·9135	$\bar{2}$·9226	$\bar{2}$·9315	16	32	49	65	81
5	$\bar{2}$·9403	$\bar{2}$·9489	$\bar{2}$·9573	$\bar{2}$·9655	$\bar{2}$·9736	$\bar{2}$·9816	$\bar{2}$·9894	$\bar{2}$·9970	$\bar{1}$·0046	$\bar{1}$·0120	13	26	40	53	66
6	$\bar{1}$·0192	$\bar{1}$·0264	$\bar{1}$·0334	$\bar{1}$·0403	$\bar{1}$·0472	$\bar{1}$·0539	$\bar{1}$·0605	$\bar{1}$·0670	$\bar{1}$·0734	$\bar{1}$·0797	11	22	33	45	56
7	$\bar{1}$·0859	$\bar{1}$·0920	$\bar{1}$·0981	$\bar{1}$·1040	$\bar{1}$·1099	$\bar{1}$·1157	$\bar{1}$·1214	$\bar{1}$·1271	$\bar{1}$·1326	$\bar{1}$·1381	10	19	29	39	48
8	$\bar{1}$·1436	$\bar{1}$·1489	$\bar{1}$·1542	$\bar{1}$·1594	$\bar{1}$·1646	$\bar{1}$·1697	$\bar{1}$·1747	$\bar{1}$·1797	$\bar{1}$·1847	$\bar{1}$·1895	9	17	25	34	42
9	$\bar{1}$·1943	$\bar{1}$·1991	$\bar{1}$·2038	$\bar{1}$·2085	$\bar{1}$·2131	$\bar{1}$·2176	$\bar{1}$·2221	$\bar{1}$·2266	$\bar{1}$·2310	$\bar{1}$·2353	8	15	23	30	38
10	$\bar{1}$·2397	$\bar{1}$·2439	$\bar{1}$·2482	$\bar{1}$·2524	$\bar{1}$·2565	$\bar{1}$·2606	$\bar{1}$·2647	$\bar{1}$·2687	$\bar{1}$·2727	$\bar{1}$·2767	7	14	21	27	34
11	$\bar{1}$·2806	$\bar{1}$·2845	$\bar{1}$·2883	$\bar{1}$·2921	$\bar{1}$·2959	$\bar{1}$·2997	$\bar{1}$·3034	$\bar{1}$·3070	$\bar{1}$·3107	$\bar{1}$·3143	6	12	19	25	31
12	$\bar{1}$·3179	$\bar{1}$·3214	$\bar{1}$·3250	$\bar{1}$·3284	$\bar{1}$·3319	$\bar{1}$·3353	$\bar{1}$·3387	$\bar{1}$·3421	$\bar{1}$·3455	$\bar{1}$·3488	6	11	17	23	29
13	$\bar{1}$·3521	$\bar{1}$·3554	$\bar{1}$·3586	$\bar{1}$·3618	$\bar{1}$·3650	$\bar{1}$·3682	$\bar{1}$·3713	$\bar{1}$·3745	$\bar{1}$·3775	$\bar{1}$·3806	5	11	16	21	26
14	$\bar{1}$·3837	$\bar{1}$·3867	$\bar{1}$·3897	$\bar{1}$·3927	$\bar{1}$·3957	$\bar{1}$·3986	$\bar{1}$·4015	$\bar{1}$·4044	$\bar{1}$·4073	$\bar{1}$·4102	5	10	15	20	24
15	$\bar{1}$·4130	$\bar{1}$·4158	$\bar{1}$·4186	$\bar{1}$·4214	$\bar{1}$·4242	$\bar{1}$·4269	$\bar{1}$·4296	$\bar{1}$·4323	$\bar{1}$·4350	$\bar{1}$·4377	5	9	14	18	23
16	$\bar{1}$·4403	$\bar{1}$·4430	$\bar{1}$·4456	$\bar{1}$·4482	$\bar{1}$·4508	$\bar{1}$·4533	$\bar{1}$·4559	$\bar{1}$·4584	$\bar{1}$·4609	$\bar{1}$·4634	4	9	13	17	21
17	$\bar{1}$·4659	$\bar{1}$·4684	$\bar{1}$·4709	$\bar{1}$·4733	$\bar{1}$·4757	$\bar{1}$·4781	$\bar{1}$·4805	$\bar{1}$·4829	$\bar{1}$·4853	$\bar{1}$·4876	4	8	12	16	20
18	$\bar{1}$·4900	$\bar{1}$·4923	$\bar{1}$·4946	$\bar{1}$·4969	$\bar{1}$·4992	$\bar{1}$·5015	$\bar{1}$·5037	$\bar{1}$·5060	$\bar{1}$·5082	$\bar{1}$·5104	4	8	11	15	19
19	$\bar{1}$·5126	$\bar{1}$·5148	$\bar{1}$·5170	$\bar{1}$·5192	$\bar{1}$·5213	$\bar{1}$·5235	$\bar{1}$·5256	$\bar{1}$·5278	$\bar{1}$·5299	$\bar{1}$·5320	4	7	11	14	18
20	$\bar{1}$·5341	$\bar{1}$·5361	$\bar{1}$·5382	$\bar{1}$·5402	$\bar{1}$·5423	$\bar{1}$·5443	$\bar{1}$·5463	$\bar{1}$·5484	$\bar{1}$·5504	$\bar{1}$·5523	3	7	10	14	17
21	$\bar{1}$·5543	$\bar{1}$·5563	$\bar{1}$·5583	$\bar{1}$·5602	$\bar{1}$·5621	$\bar{1}$·5641	$\bar{1}$·5660	$\bar{1}$·5679	$\bar{1}$·5698	$\bar{1}$·5717	3	6	10	13	16
22	$\bar{1}$·5736	$\bar{1}$·5754	$\bar{1}$·5773	$\bar{1}$·5792	$\bar{1}$·5810	$\bar{1}$·5828	$\bar{1}$·5847	$\bar{1}$·5865	$\bar{1}$·5883	$\bar{1}$·5901	3	6	9	12	15
23	$\bar{1}$·5919	$\bar{1}$·5937	$\bar{1}$·5954	$\bar{1}$·5972	$\bar{1}$·5990	$\bar{1}$·6007	$\bar{1}$·6024	$\bar{1}$·6042	$\bar{1}$·6059	$\bar{1}$·6076	3	6	9	12	15
24	$\bar{1}$·6093	$\bar{1}$·6110	$\bar{1}$·6127	$\bar{1}$·6144	$\bar{1}$·6161	$\bar{1}$·6177	$\bar{1}$·6194	$\bar{1}$·6210	$\bar{1}$·6227	$\bar{1}$·6243	3	6	8	11	14
25	$\bar{1}$·6259	$\bar{1}$·6276	$\bar{1}$·6292	$\bar{1}$·6308	$\bar{1}$·6324	$\bar{1}$·6340	$\bar{1}$·6356	$\bar{1}$·6371	$\bar{1}$·6387	$\bar{1}$·6403	3	5	8	11	13
26	$\bar{1}$·6418	$\bar{1}$·6434	$\bar{1}$·6449	$\bar{1}$·6465	$\bar{1}$·6480	$\bar{1}$·6495	$\bar{1}$·6510	$\bar{1}$·6526	$\bar{1}$·6541	$\bar{1}$·6556	3	5	8	10	13
27	$\bar{1}$·6570	$\bar{1}$·6585	$\bar{1}$·6600	$\bar{1}$·6615	$\bar{1}$·6629	$\bar{1}$·6644	$\bar{1}$·6659	$\bar{1}$·6673	$\bar{1}$·6687	$\bar{1}$·6702	2	5	7	10	12
28	$\bar{1}$·6716	$\bar{1}$·6730	$\bar{1}$·6744	$\bar{1}$·6759	$\bar{1}$·6773	$\bar{1}$·6787	$\bar{1}$·6801	$\bar{1}$·6814	$\bar{1}$·6828	$\bar{1}$·6842	2	5	7	9	12
29	$\bar{1}$·6856	$\bar{1}$·6869	$\bar{1}$·6883	$\bar{1}$·6896	$\bar{1}$·6910	$\bar{1}$·6923	$\bar{1}$·6937	$\bar{1}$·6950	$\bar{1}$·6963	$\bar{1}$·6977	2	4	7	9	11
30	$\bar{1}$·6990	$\bar{1}$·7003	$\bar{1}$·7016	$\bar{1}$·7029	$\bar{1}$·7042	$\bar{1}$·7055	$\bar{1}$·7068	$\bar{1}$·7080	$\bar{1}$·7093	$\bar{1}$·7106	2	4	6	9	11
31	$\bar{1}$·7118	$\bar{1}$·7131	$\bar{1}$·7144	$\bar{1}$·7156	$\bar{1}$·7168	$\bar{1}$·7181	$\bar{1}$·7193	$\bar{1}$·7205	$\bar{1}$·7218	$\bar{1}$·7230	2	4	6	8	10
32	$\bar{1}$·7242	$\bar{1}$·7254	$\bar{1}$·7266	$\bar{1}$·7278	$\bar{1}$·7290	$\bar{1}$·7302	$\bar{1}$·7314	$\bar{1}$·7326	$\bar{1}$·7338	$\bar{1}$·7349	2	4	6	8	10
33	$\bar{1}$·7361	$\bar{1}$·7373	$\bar{1}$·7384	$\bar{1}$·7396	$\bar{1}$·7407	$\bar{1}$·7419	$\bar{1}$·7430	$\bar{1}$·7442	$\bar{1}$·7453	$\bar{1}$·7464	2	4	6	8	10
34	$\bar{1}$·7476	$\bar{1}$·7487	$\bar{1}$·7498	$\bar{1}$·7509	$\bar{1}$·7520	$\bar{1}$·7531	$\bar{1}$·7542	$\bar{1}$·7553	$\bar{1}$·7564	$\bar{1}$·7575	2	4	6	7	9
35	$\bar{1}$·7586	$\bar{1}$·7597	$\bar{1}$·7607	$\bar{1}$·7618	$\bar{1}$·7629	$\bar{1}$·7640	$\bar{1}$·7650	$\bar{1}$·7661	$\bar{1}$·7671	$\bar{1}$·7682	2	4	5	7	9
36	$\bar{1}$·7692	$\bar{1}$·7703	$\bar{1}$·7713	$\bar{1}$·7723	$\bar{1}$·7734	$\bar{1}$·7744	$\bar{1}$·7754	$\bar{1}$·7764	$\bar{1}$·7774	$\bar{1}$·7785	2	3	5	7	9
37	$\bar{1}$·7795	$\bar{1}$·7805	$\bar{1}$·7815	$\bar{1}$·7825	$\bar{1}$·7835	$\bar{1}$·7844	$\bar{1}$·7854	$\bar{1}$·7864	$\bar{1}$·7874	$\bar{1}$·7884	2	3	5	7	8
38	$\bar{1}$·7893	$\bar{1}$·7903	$\bar{1}$·7913	$\bar{1}$·7922	$\bar{1}$·7932	$\bar{1}$·7941	$\bar{1}$·7951	$\bar{1}$·7960	$\bar{1}$·7970	$\bar{1}$·7979	2	3	5	6	8
39	$\bar{1}$·7989	$\bar{1}$·7998	$\bar{1}$·8007	$\bar{1}$·8017	$\bar{1}$·8026	$\bar{1}$·8035	$\bar{1}$·8044	$\bar{1}$·8053	$\bar{1}$·8063	$\bar{1}$·8072	2	3	5	6	8
40	$\bar{1}$·8081	$\bar{1}$·8090	$\bar{1}$·8099	$\bar{1}$·8108	$\bar{1}$·8117	$\bar{1}$·8125	$\bar{1}$·8134	$\bar{1}$·8143	$\bar{1}$·8152	$\bar{1}$·8161	1	3	4	6	7
41	$\bar{1}$·8169	$\bar{1}$·8178	$\bar{1}$·8187	$\bar{1}$·8195	$\bar{1}$·8204	$\bar{1}$·8213	$\bar{1}$·8221	$\bar{1}$·8230	$\bar{1}$·8238	$\bar{1}$·8247	1	3	4	6	7
42	$\bar{1}$·8255	$\bar{1}$·8264	$\bar{1}$·8272	$\bar{1}$·8280	$\bar{1}$·8289	$\bar{1}$·8297	$\bar{1}$·8305	$\bar{1}$·8313	$\bar{1}$·8322	$\bar{1}$·8330	1	3	4	6	7
43	$\bar{1}$·8338	$\bar{1}$·8346	$\bar{1}$·8354	$\bar{1}$·8362	$\bar{1}$·8370	$\bar{1}$·8378	$\bar{1}$·8386	$\bar{1}$·8394	$\bar{1}$·8402	$\bar{1}$·8410	1	3	4	5	7
44	$\bar{1}$·8418	$\bar{1}$·8426	$\bar{1}$·8433	$\bar{1}$·8441	$\bar{1}$·8449	$\bar{1}$·8457	$\bar{1}$·8464	$\bar{1}$·8472	$\bar{1}$·8480	$\bar{1}$·8487	1	3	4	5	6

x°	0°·0 0′	0°·1 6′	0°·2 12′	0°·3 18′	0°·4 24′	0°·5 30′	0°·6 36′	0°·7 42′	0°·8 48′	0°·9 54′	1′	2′	3′	4′	5′
45°	$\bar{1}.8495$	$\bar{1}.8502$	$\bar{1}.8510$	$\bar{1}.8517$	$\bar{1}.8525$	$\bar{1}.8532$	$\bar{1}.8540$	$\bar{1}.8547$	$\bar{1}.8555$	$\bar{1}.8562$	1	2	4	5	6
46	$\bar{1}.8569$	$\bar{1}.8577$	$\bar{1}.8584$	$\bar{1}.8591$	$\bar{1}.8598$	$\bar{1}.8606$	$\bar{1}.8613$	$\bar{1}.8620$	$\bar{1}.8627$	$\bar{1}.8634$	1	2	4	5	6
47	$\bar{1}.8641$	$\bar{1}.8648$	$\bar{1}.8655$	$\bar{1}.8662$	$\bar{1}.8669$	$\bar{1}.8676$	$\bar{1}.8683$	$\bar{1}.8690$	$\bar{1}.8697$	$\bar{1}.8704$	1	2	3	5	6
48	$\bar{1}.8711$	$\bar{1}.8718$	$\bar{1}.8724$	$\bar{1}.8731$	$\bar{1}.8738$	$\bar{1}.8745$	$\bar{1}.8751$	$\bar{1}.8758$	$\bar{1}.8765$	$\bar{1}.8771$	1	2	3	4	6
49	$\bar{1}.8778$	$\bar{1}.8784$	$\bar{1}.8791$	$\bar{1}.8797$	$\bar{1}.8804$	$\bar{1}.8810$	$\bar{1}.8817$	$\bar{1}.8823$	$\bar{1}.8830$	$\bar{1}.8836$	1	2	3	4	5
50	$\bar{1}.8843$	$\bar{1}.8849$	$\bar{1}.8855$	$\bar{1}.8862$	$\bar{1}.8868$	$\bar{1}.8874$	$\bar{1}.8880$	$\bar{1}.8887$	$\bar{1}.8893$	$\bar{1}.8899$	1	2	3	4	5
51	$\bar{1}.8905$	$\bar{1}.8911$	$\bar{1}.8917$	$\bar{1}.8923$	$\bar{1}.8929$	$\bar{1}.8935$	$\bar{1}.8941$	$\bar{1}.8947$	$\bar{1}.8953$	$\bar{1}.8959$	1	2	3	4	5
52	$\bar{1}.8965$	$\bar{1}.8971$	$\bar{1}.8977$	$\bar{1}.8983$	$\bar{1}.8989$	$\bar{1}.8995$	$\bar{1}.9000$	$\bar{1}.9006$	$\bar{1}.9012$	$\bar{1}.9018$	1	2	3	4	5
53	$\bar{1}.9023$	$\bar{1}.9029$	$\bar{1}.9035$	$\bar{1}.9041$	$\bar{1}.9046$	$\bar{1}.9052$	$\bar{1}.9057$	$\bar{1}.9063$	$\bar{1}.9069$	$\bar{1}.9074$	1	2	3	4	5
54	$\bar{1}.9080$	$\bar{1}.9085$	$\bar{1}.9091$	$\bar{1}.9096$	$\bar{1}.9101$	$\bar{1}.9107$	$\bar{1}.9112$	$\bar{1}.9118$	$\bar{1}.9123$	$\bar{1}.9128$	1	2	3	4	5
55	$\bar{1}.9134$	$\bar{1}.9139$	$\bar{1}.9144$	$\bar{1}.9149$	$\bar{1}.9155$	$\bar{1}.9160$	$\bar{1}.9165$	$\bar{1}.9170$	$\bar{1}.9175$	$\bar{1}.9181$	1	2	3	3	4
56	$\bar{1}.9186$	$\bar{1}.9191$	$\bar{1}.9196$	$\bar{1}.9201$	$\bar{1}.9206$	$\bar{1}.9211$	$\bar{1}.9216$	$\bar{1}.9221$	$\bar{1}.9226$	$\bar{1}.9231$	1	2	3	3	4
57	$\bar{1}.9236$	$\bar{1}.9241$	$\bar{1}.9246$	$\bar{1}.9251$	$\bar{1}.9255$	$\bar{1}.9260$	$\bar{1}.9265$	$\bar{1}.9270$	$\bar{1}.9275$	$\bar{1}.9279$	1	2	2	3	4
58	$\bar{1}.9284$	$\bar{1}.9289$	$\bar{1}.9294$	$\bar{1}.9298$	$\bar{1}.9303$	$\bar{1}.9308$	$\bar{1}.9312$	$\bar{1}.9317$	$\bar{1}.9322$	$\bar{1}.9326$	1	2	2	3	4
59	$\bar{1}.9331$	$\bar{1}.9335$	$\bar{1}.9340$	$\bar{1}.9344$	$\bar{1}.9349$	$\bar{1}.9353$	$\bar{1}.9358$	$\bar{1}.9362$	$\bar{1}.9367$	$\bar{1}.9371$	1	1	2	3	4
60	$\bar{1}.9375$	$\bar{1}.9380$	$\bar{1}.9384$	$\bar{1}.9388$	$\bar{1}.9393$	$\bar{1}.9397$	$\bar{1}.9401$	$\bar{1}.9406$	$\bar{1}.9410$	$\bar{1}.9414$	1	1	2	3	4
61	$\bar{1}.9418$	$\bar{1}.9422$	$\bar{1}.9427$	$\bar{1}.9431$	$\bar{1}.9435$	$\bar{1}.9439$	$\bar{1}.9443$	$\bar{1}.9447$	$\bar{1}.9451$	$\bar{1}.9455$	1	1	2	3	3
62	$\bar{1}.9459$	$\bar{1}.9463$	$\bar{1}.9467$	$\bar{1}.9471$	$\bar{1}.9475$	$\bar{1}.9479$	$\bar{1}.9483$	$\bar{1}.9487$	$\bar{1}.9491$	$\bar{1}.9495$	1	1	2	3	3
63	$\bar{1}.9499$	$\bar{1}.9503$	$\bar{1}.9506$	$\bar{1}.9510$	$\bar{1}.9514$	$\bar{1}.9518$	$\bar{1}.9522$	$\bar{1}.9525$	$\bar{1}.9529$	$\bar{1}.9533$	1	1	2	3	3
64	$\bar{1}.9537$	$\bar{1}.9540$	$\bar{1}.9544$	$\bar{1}.9548$	$\bar{1}.9551$	$\bar{1}.9555$	$\bar{1}.9558$	$\bar{1}.9562$	$\bar{1}.9566$	$\bar{1}.9569$	1	1	2	2	3
65	$\bar{1}.9573$	$\bar{1}.9576$	$\bar{1}.9580$	$\bar{1}.9583$	$\bar{1}.9587$	$\bar{1}.9590$	$\bar{1}.9594$	$\bar{1}.9597$	$\bar{1}.9601$	$\bar{1}.9604$	1	1	2	2	3
66	$\bar{1}.9607$	$\bar{1}.9611$	$\bar{1}.9614$	$\bar{1}.9617$	$\bar{1}.9621$	$\bar{1}.9624$	$\bar{1}.9627$	$\bar{1}.9631$	$\bar{1}.9634$	$\bar{1}.9637$	1	1	2	2	3
67	$\bar{1}.9640$	$\bar{1}.9643$	$\bar{1}.9647$	$\bar{1}.9650$	$\bar{1}.9653$	$\bar{1}.9656$	$\bar{1}.9659$	$\bar{1}.9662$	$\bar{1}.9666$	$\bar{1}.9669$	1	1	2	2	3
68	$\bar{1}.9672$	$\bar{1}.9675$	$\bar{1}.9678$	$\bar{1}.9681$	$\bar{1}.9684$	$\bar{1}.9687$	$\bar{1}.9690$	$\bar{1}.9693$	$\bar{1}.9696$	$\bar{1}.9699$	0	1	1	2	2
69	$\bar{1}.9702$	$\bar{1}.9704$	$\bar{1}.9707$	$\bar{1}.9710$	$\bar{1}.9713$	$\bar{1}.9716$	$\bar{1}.9719$	$\bar{1}.9722$	$\bar{1}.9724$	$\bar{1}.9727$	0	1	1	2	2
70	$\bar{1}.9730$	$\bar{1}.9733$	$\bar{1}.9735$	$\bar{1}.9738$	$\bar{1}.9741$	$\bar{1}.9743$	$\bar{1}.9746$	$\bar{1}.9749$	$\bar{1}.9751$	$\bar{1}.9754$	0	1	1	2	2
71	$\bar{1}.9757$	$\bar{1}.9759$	$\bar{1}.9762$	$\bar{1}.9764$	$\bar{1}.9767$	$\bar{1}.9770$	$\bar{1}.9772$	$\bar{1}.9775$	$\bar{1}.9777$	$\bar{1}.9780$	0	1	1	2	2
72	$\bar{1}.9782$	$\bar{1}.9785$	$\bar{1}.9787$	$\bar{1}.9789$	$\bar{1}.9792$	$\bar{1}.9794$	$\bar{1}.9797$	$\bar{1}.9799$	$\bar{1}.9801$	$\bar{1}.9804$	0	1	1	2	2
73	$\bar{1}.9806$	$\bar{1}.9808$	$\bar{1}.9811$	$\bar{1}.9813$	$\bar{1}.9815$	$\bar{1}.9817$	$\bar{1}.9820$	$\bar{1}.9822$	$\bar{1}.9824$	$\bar{1}.9826$	0	1	1	1	2
74	$\bar{1}.9828$	$\bar{1}.9831$	$\bar{1}.9833$	$\bar{1}.9835$	$\bar{1}.9837$	$\bar{1}.9839$	$\bar{1}.9841$	$\bar{1}.9843$	$\bar{1}.9845$	$\bar{1}.9847$	0	1	1	1	2
75	$\bar{1}.9849$	$\bar{1}.9851$	$\bar{1}.9853$	$\bar{1}.9855$	$\bar{1}.9857$	$\bar{1}.9859$	$\bar{1}.9861$	$\bar{1}.9863$	$\bar{1}.9865$	$\bar{1}.9867$	0	1	1	1	2
76	$\bar{1}.9869$	$\bar{1}.9871$	$\bar{1}.9873$	$\bar{1}.9875$	$\bar{1}.9876$	$\bar{1}.9878$	$\bar{1}.9880$	$\bar{1}.9882$	$\bar{1}.9884$	$\bar{1}.9885$	0	1	1	1	2
77	$\bar{1}.9887$	$\bar{1}.9889$	$\bar{1}.9891$	$\bar{1}.9892$	$\bar{1}.9894$	$\bar{1}.9896$	$\bar{1}.9897$	$\bar{1}.9899$	$\bar{1}.9901$	$\bar{1}.9902$	0	1	1	1	1
78	$\bar{1}.9904$	$\bar{1}.9906$	$\bar{1}.9907$	$\bar{1}.9909$	$\bar{1}.9910$	$\bar{1}.9912$	$\bar{1}.9913$	$\bar{1}.9915$	$\bar{1}.9916$	$\bar{1}.9918$	0	1	1	1	1
79	$\bar{1}.9919$	$\bar{1}.9921$	$\bar{1}.9922$	$\bar{1}.9924$	$\bar{1}.9925$	$\bar{1}.9927$	$\bar{1}.9928$	$\bar{1}.9929$	$\bar{1}.9931$	$\bar{1}.9932$	0	0	1	1	1
80	$\bar{1}.9934$	$\bar{1}.9935$	$\bar{1}.9936$	$\bar{1}.9937$	$\bar{1}.9939$	$\bar{1}.9940$	$\bar{1}.9941$	$\bar{1}.9943$	$\bar{1}.9944$	$\bar{1}.9945$	0	0	1	1	1
81	$\bar{1}.9946$	$\bar{1}.9947$	$\bar{1}.9949$	$\bar{1}.9950$	$\bar{1}.9951$	$\bar{1}.9952$	$\bar{1}.9953$	$\bar{1}.9954$	$\bar{1}.9955$	$\bar{1}.9956$	0	0	1	1	1
82	$\bar{1}.9958$	$\bar{1}.9959$	$\bar{1}.9960$	$\bar{1}.9961$	$\bar{1}.9962$	$\bar{1}.9963$	$\bar{1}.9964$	$\bar{1}.9965$	$\bar{1}.9966$	$\bar{1}.9967$	0	0	1	1	1
83	$\bar{1}.9968$	$\bar{1}.9968$	$\bar{1}.9969$	$\bar{1}.9970$	$\bar{1}.9971$	$\bar{1}.9972$	$\bar{1}.9973$	$\bar{1}.9974$	$\bar{1}.9975$	$\bar{1}.9975$	0	0	0	1	1
84	$\bar{1}.9976$	$\bar{1}.9977$	$\bar{1}.9978$	$\bar{1}.9978$	$\bar{1}.9979$	$\bar{1}.9980$	$\bar{1}.9981$	$\bar{1}.9981$	$\bar{1}.9982$	$\bar{1}.9983$	0	0	0	0	1
85	$\bar{1}.9983$	$\bar{1}.9984$	$\bar{1}.9985$	$\bar{1}.9985$	$\bar{1}.9986$	$\bar{1}.9987$	$\bar{1}.9987$	$\bar{1}.9988$	$\bar{1}.9988$	$\bar{1}.9989$	0	0	0	0	0
86	$\bar{1}.9989$	$\bar{1}.9990$	$\bar{1}.9990$	$\bar{1}.9991$	$\bar{1}.9991$	$\bar{1}.9992$	$\bar{1}.9992$	$\bar{1}.9993$	$\bar{1}.9993$	$\bar{1}.9994$	0	0	0	0	0
87	$\bar{1}.9994$	$\bar{1}.9994$	$\bar{1}.9995$	$\bar{1}.9995$	$\bar{1}.9996$	$\bar{1}.9996$	$\bar{1}.9996$	$\bar{1}.9996$	$\bar{1}.9997$	$\bar{1}.9997$	0	0	0	0	0
88	$\bar{1}.9997$	$\bar{1}.9998$	$\bar{1}.9998$	$\bar{1}.9998$	$\bar{1}.9998$	$\bar{1}.9999$	$\bar{1}.9999$	$\bar{1}.9999$	$\bar{1}.9999$	$\bar{1}.9999$	0	0	0	0	0
89	$\bar{1}.9999$	$\bar{1}.9999$	0.0000	0.0000	0.0000	0.0000	0.0000	0.0000	0.0000	0.0000	0	0	0	0	0

Table 6. Logarithms of Cosines $\log_{10} \cos x^\circ$

x°	0°·0 0′	0°·1 6′	0°·2 12′	0°·3 18′	0°·4 24′	0°·5 30′	0°·6 36′	0°·7 42′	0°·8 48′	0°·9 54′	SUBTRACT 1′	2′	3′	4′	5′
0°	0·0000	0·0000	0·0000	0·0000	0·0000	0·0000	0·0000	0·0000	0·0000	$\bar{1}$·9999	0	0	0	0	0
1	$\bar{1}$·9999	$\bar{1}$·9999	$\bar{1}$·9999	$\bar{1}$·9999	$\bar{1}$·9999	$\bar{1}$·9999	$\bar{1}$·9998	$\bar{1}$·9998	$\bar{1}$·9998	$\bar{1}$·9998	0	0	0	0	0
2	$\bar{1}$·9997	$\bar{1}$·9997	$\bar{1}$·9997	$\bar{1}$·9996	$\bar{1}$·9996	$\bar{1}$·9996	$\bar{1}$·9996	$\bar{1}$·9995	$\bar{1}$·9995	$\bar{1}$·9994	0	0	0	0	0
3	$\bar{1}$·9994	$\bar{1}$·9994	$\bar{1}$·9993	$\bar{1}$·9993	$\bar{1}$·9992	$\bar{1}$·9992	$\bar{1}$·9991	$\bar{1}$·9991	$\bar{1}$·9990	$\bar{1}$·9990	0	0	0	0	0
4	$\bar{1}$·9989	$\bar{1}$·9989	$\bar{1}$·9988	$\bar{1}$·9988	$\bar{1}$·9987	$\bar{1}$·9987	$\bar{1}$·9986	$\bar{1}$·9985	$\bar{1}$·9985	$\bar{1}$·9984	0	0	0	0	0
5	$\bar{1}$·9983	$\bar{1}$·9983	$\bar{1}$·9982	$\bar{1}$·9981	$\bar{1}$·9981	$\bar{1}$·9980	$\bar{1}$·9979	$\bar{1}$·9978	$\bar{1}$·9978	$\bar{1}$·9977	0	0	0	0	1
6	$\bar{1}$·9976	$\bar{1}$·9975	$\bar{1}$·9975	$\bar{1}$·9974	$\bar{1}$·9973	$\bar{1}$·9972	$\bar{1}$·9971	$\bar{1}$·9970	$\bar{1}$·9969	$\bar{1}$·9968	0	0	0	1	1
7	$\bar{1}$·9968	$\bar{1}$·9967	$\bar{1}$·9966	$\bar{1}$·9965	$\bar{1}$·9964	$\bar{1}$·9963	$\bar{1}$·9962	$\bar{1}$·9961	$\bar{1}$·9960	$\bar{1}$·9959	0	0	0	1	1
8	$\bar{1}$·9958	$\bar{1}$·9956	$\bar{1}$·9955	$\bar{1}$·9954	$\bar{1}$·9953	$\bar{1}$·9952	$\bar{1}$·9951	$\bar{1}$·9950	$\bar{1}$·9949	$\bar{1}$·9947	0	0	1	1	1
9	$\bar{1}$·9946	$\bar{1}$·9945	$\bar{1}$·9944	$\bar{1}$·9943	$\bar{1}$·9941	$\bar{1}$·9940	$\bar{1}$·9939	$\bar{1}$·9937	$\bar{1}$·9936	$\bar{1}$·9935	0	0	1	1	1
10	$\bar{1}$·9934	$\bar{1}$·9932	$\bar{1}$·9931	$\bar{1}$·9929	$\bar{1}$·9928	$\bar{1}$·9927	$\bar{1}$·9925	$\bar{1}$·9924	$\bar{1}$·9922	$\bar{1}$·9921	0	0	1	1	1
11	$\bar{1}$·9919	$\bar{1}$·9918	$\bar{1}$·9916	$\bar{1}$·9915	$\bar{1}$·9913	$\bar{1}$·9912	$\bar{1}$·9910	$\bar{1}$·9909	$\bar{1}$·9907	$\bar{1}$·9906	0	1	1	1	1
12	$\bar{1}$·9904	$\bar{1}$·9902	$\bar{1}$·9901	$\bar{1}$·9899	$\bar{1}$·9897	$\bar{1}$·9896	$\bar{1}$·9894	$\bar{1}$·9892	$\bar{1}$·9891	$\bar{1}$·9889	0	1	1	1	1
13	$\bar{1}$·9887	$\bar{1}$·9885	$\bar{1}$·9884	$\bar{1}$·9882	$\bar{1}$·9880	$\bar{1}$·9878	$\bar{1}$·9876	$\bar{1}$·9875	$\bar{1}$·9873	$\bar{1}$·9871	0	1	1	1	2
14	$\bar{1}$·9869	$\bar{1}$·9867	$\bar{1}$·9865	$\bar{1}$·9863	$\bar{1}$·9861	$\bar{1}$·9859	$\bar{1}$·9857	$\bar{1}$·9855	$\bar{1}$·9853	$\bar{1}$·9851	0	1	1	1	2
15	$\bar{1}$·9849	$\bar{1}$·9847	$\bar{1}$·9845	$\bar{1}$·9843	$\bar{1}$·9841	$\bar{1}$·9839	$\bar{1}$·9837	$\bar{1}$·9835	$\bar{1}$·9833	$\bar{1}$·9831	0	1	1	1	2
16	$\bar{1}$·9828	$\bar{1}$·9826	$\bar{1}$·9824	$\bar{1}$·9822	$\bar{1}$·9820	$\bar{1}$·9817	$\bar{1}$·9815	$\bar{1}$·9813	$\bar{1}$·9811	$\bar{1}$·9808	0	1	1	1	2
17	$\bar{1}$·9806	$\bar{1}$·9804	$\bar{1}$·9801	$\bar{1}$·9799	$\bar{1}$·9797	$\bar{1}$·9794	$\bar{1}$·9792	$\bar{1}$·9789	$\bar{1}$·9787	$\bar{1}$·9785	0	1	1	2	2
18	$\bar{1}$·9782	$\bar{1}$·9780	$\bar{1}$·9777	$\bar{1}$·9775	$\bar{1}$·9772	$\bar{1}$·9770	$\bar{1}$·9767	$\bar{1}$·9764	$\bar{1}$·9762	$\bar{1}$·9759	0	1	1	2	2
19	$\bar{1}$·9757	$\bar{1}$·9754	$\bar{1}$·9751	$\bar{1}$·9749	$\bar{1}$·9746	$\bar{1}$·9743	$\bar{1}$·9741	$\bar{1}$·9738	$\bar{1}$·9735	$\bar{1}$·9733	0	1	1	2	2
20	$\bar{1}$·9730	$\bar{1}$·9727	$\bar{1}$·9724	$\bar{1}$·9722	$\bar{1}$·9719	$\bar{1}$·9716	$\bar{1}$·9713	$\bar{1}$·9710	$\bar{1}$·9707	$\bar{1}$·9704	0	1	1	2	2
21	$\bar{1}$·9702	$\bar{1}$·9699	$\bar{1}$·9696	$\bar{1}$·9693	$\bar{1}$·9690	$\bar{1}$·9687	$\bar{1}$·9684	$\bar{1}$·9681	$\bar{1}$·9678	$\bar{1}$·9675	0	1	1	2	2
22	$\bar{1}$·9672	$\bar{1}$·9669	$\bar{1}$·9666	$\bar{1}$·9662	$\bar{1}$·9659	$\bar{1}$·9656	$\bar{1}$·9653	$\bar{1}$·9650	$\bar{1}$·9647	$\bar{1}$·9643	1	1	2	2	3
23	$\bar{1}$·9640	$\bar{1}$·9637	$\bar{1}$·9634	$\bar{1}$·9631	$\bar{1}$·9627	$\bar{1}$·9624	$\bar{1}$·9621	$\bar{1}$·9617	$\bar{1}$·9614	$\bar{1}$·9611	1	1	2	2	3
24	$\bar{1}$·9607	$\bar{1}$·9604	$\bar{1}$·9601	$\bar{1}$·9597	$\bar{1}$·9594	$\bar{1}$·9590	$\bar{1}$·9587	$\bar{1}$·9583	$\bar{1}$·9580	$\bar{1}$·9576	1	1	2	2	3
25	$\bar{1}$·9573	$\bar{1}$·9569	$\bar{1}$·9566	$\bar{1}$·9562	$\bar{1}$·9558	$\bar{1}$·9555	$\bar{1}$·9551	$\bar{1}$·9548	$\bar{1}$·9544	$\bar{1}$·9540	1	1	2	2	3
26	$\bar{1}$·9537	$\bar{1}$·9533	$\bar{1}$·9529	$\bar{1}$·9525	$\bar{1}$·9522	$\bar{1}$·9518	$\bar{1}$·9514	$\bar{1}$·9510	$\bar{1}$·9506	$\bar{1}$·9503	1	1	2	3	3
27	$\bar{1}$·9499	$\bar{1}$·9495	$\bar{1}$·9491	$\bar{1}$·9487	$\bar{1}$·9483	$\bar{1}$·9479	$\bar{1}$·9475	$\bar{1}$·9471	$\bar{1}$·9467	$\bar{1}$·9463	1	1	2	3	3
28	$\bar{1}$·9459	$\bar{1}$·9455	$\bar{1}$·9451	$\bar{1}$·9447	$\bar{1}$·9443	$\bar{1}$·9439	$\bar{1}$·9435	$\bar{1}$·9431	$\bar{1}$·9427	$\bar{1}$·9422	1	1	2	3	3
29	$\bar{1}$·9418	$\bar{1}$·9414	$\bar{1}$·9410	$\bar{1}$·9406	$\bar{1}$·9401	$\bar{1}$·9397	$\bar{1}$·9393	$\bar{1}$·9388	$\bar{1}$·9384	$\bar{1}$·9380	1	1	2	3	4
30	$\bar{1}$·9375	$\bar{1}$·9371	$\bar{1}$·9367	$\bar{1}$·9362	$\bar{1}$·9358	$\bar{1}$·9353	$\bar{1}$·9349	$\bar{1}$·9344	$\bar{1}$·9340	$\bar{1}$·9335	1	1	2	3	4
31	$\bar{1}$·9331	$\bar{1}$·9326	$\bar{1}$·9322	$\bar{1}$·9317	$\bar{1}$·9312	$\bar{1}$·9308	$\bar{1}$·9303	$\bar{1}$·9298	$\bar{1}$·9294	$\bar{1}$·9289	1	2	2	3	4
32	$\bar{1}$·9284	$\bar{1}$·9279	$\bar{1}$·9275	$\bar{1}$·9270	$\bar{1}$·9265	$\bar{1}$·9260	$\bar{1}$·9255	$\bar{1}$·9251	$\bar{1}$·9246	$\bar{1}$·9241	1	2	2	3	4
33	$\bar{1}$·9236	$\bar{1}$·9231	$\bar{1}$·9226	$\bar{1}$·9221	$\bar{1}$·9216	$\bar{1}$·9211	$\bar{1}$·9206	$\bar{1}$·9201	$\bar{1}$·9196	$\bar{1}$·9191	1	2	3	3	4
34	$\bar{1}$·9186	$\bar{1}$·9181	$\bar{1}$·9175	$\bar{1}$·9170	$\bar{1}$·9165	$\bar{1}$·9160	$\bar{1}$·9155	$\bar{1}$·9149	$\bar{1}$·9144	$\bar{1}$·9139	1	2	3	3	4
35	$\bar{1}$·9134	$\bar{1}$·9128	$\bar{1}$·9123	$\bar{1}$·9118	$\bar{1}$·9112	$\bar{1}$·9107	$\bar{1}$·9101	$\bar{1}$·9096	$\bar{1}$·9091	$\bar{1}$·9085	1	2	3	4	5
36	$\bar{1}$·9080	$\bar{1}$·9074	$\bar{1}$·9069	$\bar{1}$·9063	$\bar{1}$·9057	$\bar{1}$·9052	$\bar{1}$·9046	$\bar{1}$·9041	$\bar{1}$·9035	$\bar{1}$·9029	1	2	3	4	5
37	$\bar{1}$·9023	$\bar{1}$·9018	$\bar{1}$·9012	$\bar{1}$·9006	$\bar{1}$·9000	$\bar{1}$·8995	$\bar{1}$·8989	$\bar{1}$·8983	$\bar{1}$·8977	$\bar{1}$·8971	1	2	3	4	5
38	$\bar{1}$·8965	$\bar{1}$·8959	$\bar{1}$·8953	$\bar{1}$·8947	$\bar{1}$·8941	$\bar{1}$·8935	$\bar{1}$·8929	$\bar{1}$·8923	$\bar{1}$·8917	$\bar{1}$·8911	1	2	3	4	5
39	$\bar{1}$·8905	$\bar{1}$·8899	$\bar{1}$·8893	$\bar{1}$·8887	$\bar{1}$·8880	$\bar{1}$·8874	$\bar{1}$·8868	$\bar{1}$·8862	$\bar{1}$·8855	$\bar{1}$·8849	1	2	3	4	5
40	$\bar{1}$·8843	$\bar{1}$·8836	$\bar{1}$·8830	$\bar{1}$·8823	$\bar{1}$·8817	$\bar{1}$·8810	$\bar{1}$·8804	$\bar{1}$·8797	$\bar{1}$·8791	$\bar{1}$·8784	1	2	3	4	5
41	$\bar{1}$·8778	$\bar{1}$·8771	$\bar{1}$·8765	$\bar{1}$·8758	$\bar{1}$·8751	$\bar{1}$·8745	$\bar{1}$·8738	$\bar{1}$·8731	$\bar{1}$·8724	$\bar{1}$·8718	1	2	3	4	6
42	$\bar{1}$·8711	$\bar{1}$·8704	$\bar{1}$·8697	$\bar{1}$·8690	$\bar{1}$·8683	$\bar{1}$·8676	$\bar{1}$·8669	$\bar{1}$·8662	$\bar{1}$·8655	$\bar{1}$·8648	1	2	3	5	6
43	$\bar{1}$·8641	$\bar{1}$·8634	$\bar{1}$·8627	$\bar{1}$·8620	$\bar{1}$·8613	$\bar{1}$·8606	$\bar{1}$·8598	$\bar{1}$·8591	$\bar{1}$·8584	$\bar{1}$·8577	1	2	4	5	6
44	$\bar{1}$·8569	$\bar{1}$·8562	$\bar{1}$·8555	$\bar{1}$·8547	$\bar{1}$·8540	$\bar{1}$·8532	$\bar{1}$·8525	$\bar{1}$·8517	$\bar{1}$·8510	$\bar{1}$·8502	1	2	4	5	6

x°	0°·0 0′	0°·1 6′	0°·2 12′	0°·3 18′	0°·4 24′	0°·5 30′	0°·6 36′	0°·7 42′	0°·8 48′	0°·9 54′	SUBTRACT 1′	2′	3′	4′	5′
45	$\bar{1}$·8495	$\bar{1}$·8487	$\bar{1}$·8480	$\bar{1}$·8472	$\bar{1}$·8464	$\bar{1}$·8457	$\bar{1}$·8449	$\bar{1}$·8441	$\bar{1}$·8433	$\bar{1}$·8426	1	3	4	5	6
46	$\bar{1}$·8418	$\bar{1}$·8410	$\bar{1}$·8402	$\bar{1}$·8394	$\bar{1}$·8386	$\bar{1}$·8378	$\bar{1}$·8370	$\bar{1}$·8362	$\bar{1}$·8354	$\bar{1}$·8346	1	3	4	5	7
47	$\bar{1}$·8338	$\bar{1}$·8330	$\bar{1}$·8322	$\bar{1}$·8313	$\bar{1}$·8305	$\bar{1}$·8297	$\bar{1}$·8289	$\bar{1}$·8280	$\bar{1}$·8272	$\bar{1}$·8264	1	3	4	6	7
48	$\bar{1}$·8255	$\bar{1}$·8247	$\bar{1}$·8238	$\bar{1}$·8230	$\bar{1}$·8221	$\bar{1}$·8213	$\bar{1}$·8204	$\bar{1}$·8195	$\bar{1}$·8187	$\bar{1}$·8178	1	3	4	6	7
49	$\bar{1}$·8169	$\bar{1}$·8161	$\bar{1}$·8152	$\bar{1}$·8143	$\bar{1}$·8134	$\bar{1}$·8125	$\bar{1}$·8117	$\bar{1}$·8108	$\bar{1}$·8099	$\bar{1}$·8090	1	3	4	6	7
50	$\bar{1}$·8081	$\bar{1}$·8072	$\bar{1}$·8063	$\bar{1}$·8053	$\bar{1}$·8044	$\bar{1}$·8035	$\bar{1}$·8026	$\bar{1}$·8017	$\bar{1}$·8007	$\bar{1}$·7998	2	3	5	6	8
51	$\bar{1}$·7989	$\bar{1}$·7979	$\bar{1}$·7970	$\bar{1}$·7960	$\bar{1}$·7951	$\bar{1}$·7941	$\bar{1}$·7932	$\bar{1}$·7922	$\bar{1}$·7913	$\bar{1}$·7903	2	3	5	6	8
52	$\bar{1}$·7893	$\bar{1}$·7884	$\bar{1}$·7874	$\bar{1}$·7864	$\bar{1}$·7854	$\bar{1}$·7844	$\bar{1}$·7835	$\bar{1}$·7825	$\bar{1}$·7815	$\bar{1}$·7805	2	3	5	7	8
53	$\bar{1}$·7795	$\bar{1}$·7785	$\bar{1}$·7774	$\bar{1}$·7764	$\bar{1}$·7754	$\bar{1}$·7744	$\bar{1}$·7734	$\bar{1}$·7723	$\bar{1}$·7713	$\bar{1}$·7703	2	3	5	7	9
54	$\bar{1}$·7692	$\bar{1}$·7682	$\bar{1}$·7671	$\bar{1}$·7661	$\bar{1}$·7650	$\bar{1}$·7640	$\bar{1}$·7629	$\bar{1}$·7618	$\bar{1}$·7607	$\bar{1}$·7597	2	4	5	7	9
55	$\bar{1}$·7586	$\bar{1}$·7575	$\bar{1}$·7564	$\bar{1}$·7553	$\bar{1}$·7542	$\bar{1}$·7531	$\bar{1}$·7520	$\bar{1}$·7509	$\bar{1}$·7498	$\bar{1}$·7487	2	4	6	7	9
56	$\bar{1}$·7476	$\bar{1}$·7464	$\bar{1}$·7453	$\bar{1}$·7442	$\bar{1}$·7430	$\bar{1}$·7419	$\bar{1}$·7407	$\bar{1}$·7396	$\bar{1}$·7384	$\bar{1}$·7373	2	4	6	8	10
57	$\bar{1}$·7361	$\bar{1}$·7349	$\bar{1}$·7338	$\bar{1}$·7326	$\bar{1}$·7314	$\bar{1}$·7302	$\bar{1}$·7290	$\bar{1}$·7278	$\bar{1}$·7266	$\bar{1}$·7254	2	4	6	8	10
58	$\bar{1}$·7242	$\bar{1}$·7230	$\bar{1}$·7218	$\bar{1}$·7205	$\bar{1}$·7193	$\bar{1}$·7181	$\bar{1}$·7168	$\bar{1}$·7156	$\bar{1}$·7144	$\bar{1}$·7131	2	4	6	8	10
59	$\bar{1}$·7118	$\bar{1}$·7106	$\bar{1}$·7093	$\bar{1}$·7080	$\bar{1}$·7068	$\bar{1}$·7055	$\bar{1}$·7042	$\bar{1}$·7029	$\bar{1}$·7016	$\bar{1}$·7003	2	4	6	9	11
60	$\bar{1}$·6990	$\bar{1}$·6977	$\bar{1}$·6963	$\bar{1}$·6950	$\bar{1}$·6937	$\bar{1}$·6923	$\bar{1}$·6910	$\bar{1}$·6896	$\bar{1}$·6883	$\bar{1}$·6869	2	4	7	9	11
61	$\bar{1}$·6856	$\bar{1}$·6842	$\bar{1}$·6828	$\bar{1}$·6814	$\bar{1}$·6801	$\bar{1}$·6787	$\bar{1}$·6773	$\bar{1}$·6759	$\bar{1}$·6744	$\bar{1}$·6730	2	5	7	9	12
62	$\bar{1}$·6716	$\bar{1}$·6702	$\bar{1}$·6687	$\bar{1}$·6673	$\bar{1}$·6659	$\bar{1}$·6644	$\bar{1}$·6629	$\bar{1}$·6615	$\bar{1}$·6600	$\bar{1}$·6585	2	5	7	10	12
63	$\bar{1}$·6570	$\bar{1}$·6556	$\bar{1}$·6541	$\bar{1}$·6526	$\bar{1}$·6510	$\bar{1}$·6495	$\bar{1}$·6480	$\bar{1}$·6465	$\bar{1}$·6449	$\bar{1}$·6434	3	5	8	10	13
64	$\bar{1}$·6418	$\bar{1}$·6403	$\bar{1}$·6387	$\bar{1}$·6371	$\bar{1}$·6356	$\bar{1}$·6340	$\bar{1}$·6324	$\bar{1}$·6308	$\bar{1}$·6292	$\bar{1}$·6276	3	5	8	11	13
65	$\bar{1}$·6259	$\bar{1}$·6243	$\bar{1}$·6227	$\bar{1}$·6210	$\bar{1}$·6194	$\bar{1}$·6177	$\bar{1}$·6161	$\bar{1}$·6144	$\bar{1}$·6127	$\bar{1}$·6110	3	6	8	11	14
66	$\bar{1}$·6093	$\bar{1}$·6076	$\bar{1}$·6059	$\bar{1}$·6042	$\bar{1}$·6024	$\bar{1}$·6007	$\bar{1}$·5990	$\bar{1}$·5972	$\bar{1}$·5954	$\bar{1}$·5937	3	6	9	12	15
67	$\bar{1}$·5919	$\bar{1}$·5901	$\bar{1}$·5883	$\bar{1}$·5865	$\bar{1}$·5847	$\bar{1}$·5828	$\bar{1}$·5810	$\bar{1}$·5792	$\bar{1}$·5773	$\bar{1}$·5754	3	6	9	12	15
68	$\bar{1}$·5736	$\bar{1}$·5717	$\bar{1}$·5698	$\bar{1}$·5679	$\bar{1}$·5660	$\bar{1}$·5641	$\bar{1}$·5621	$\bar{1}$·5602	$\bar{1}$·5583	$\bar{1}$·5563	3	6	10	13	16
69	$\bar{1}$·5543	$\bar{1}$·5523	$\bar{1}$·5504	$\bar{1}$·5484	$\bar{1}$·5463	$\bar{1}$·5443	$\bar{1}$·5423	$\bar{1}$·5402	$\bar{1}$·5382	$\bar{1}$·5361	3	7	10	14	17
70	$\bar{1}$·5341	$\bar{1}$·5320	$\bar{1}$·5299	$\bar{1}$·5278	$\bar{1}$·5256	$\bar{1}$·5235	$\bar{1}$·5213	$\bar{1}$·5192	$\bar{1}$·5170	$\bar{1}$·5148	4	7	11	14	18
71	$\bar{1}$·5126	$\bar{1}$·5104	$\bar{1}$·5082	$\bar{1}$·5060	$\bar{1}$·5037	$\bar{1}$·5015	$\bar{1}$·4992	$\bar{1}$·4969	$\bar{1}$·4946	$\bar{1}$·4923	4	8	11	15	19
72	$\bar{1}$·4900	$\bar{1}$·4876	$\bar{1}$·4853	$\bar{1}$·4829	$\bar{1}$·4805	$\bar{1}$·4781	$\bar{1}$·4757	$\bar{1}$·4733	$\bar{1}$·4709	$\bar{1}$·4684	4	8	12	16	20
73	$\bar{1}$·4659	$\bar{1}$·4634	$\bar{1}$·4609	$\bar{1}$·4584	$\bar{1}$·4559	$\bar{1}$·4533	$\bar{1}$·4508	$\bar{1}$·4482	$\bar{1}$·4456	$\bar{1}$·4430	4	9	13	17	21
74	$\bar{1}$·4403	$\bar{1}$·4377	$\bar{1}$·4350	$\bar{1}$·4323	$\bar{1}$·4296	$\bar{1}$·4269	$\bar{1}$·4242	$\bar{1}$·4214	$\bar{1}$·4186	$\bar{1}$·4158	5	9	14	18	23
75	$\bar{1}$·4130	$\bar{1}$·4102	$\bar{1}$·4073	$\bar{1}$·4044	$\bar{1}$·4015	$\bar{1}$·3986	$\bar{1}$·3957	$\bar{1}$·3927	$\bar{1}$·3897	$\bar{1}$·3867	5	10	15	20	24
76	$\bar{1}$·3837	$\bar{1}$·3806	$\bar{1}$·3775	$\bar{1}$·3745	$\bar{1}$·3713	$\bar{1}$·3682	$\bar{1}$·3650	$\bar{1}$·3618	$\bar{1}$·3586	$\bar{1}$·3554	5	11	16	21	26
77	$\bar{1}$·3521	$\bar{1}$·3488	$\bar{1}$·3455	$\bar{1}$·3421	$\bar{1}$·3387	$\bar{1}$·3353	$\bar{1}$·3319	$\bar{1}$·3284	$\bar{1}$·3250	$\bar{1}$·3214	6	11	17	23	28
78	$\bar{1}$·3179	$\bar{1}$·3143	$\bar{1}$·3107	$\bar{1}$·3070	$\bar{1}$·3034	$\bar{1}$·2997	$\bar{1}$·2959	$\bar{1}$·2921	$\bar{1}$·2883	$\bar{1}$·2845	6	12	19	25	31
79	$\bar{1}$·2806	$\bar{1}$·2767	$\bar{1}$·2727	$\bar{1}$·2687	$\bar{1}$·2647	$\bar{1}$·2606	$\bar{1}$·2565	$\bar{1}$·2524	$\bar{1}$·2482	$\bar{1}$·2439	7	14	20	27	34
80	$\bar{1}$·2397	$\bar{1}$·2353	$\bar{1}$·2310	$\bar{1}$·2266	$\bar{1}$·2221	$\bar{1}$·2176	$\bar{1}$·2131	$\bar{1}$·2085	$\bar{1}$·2038	$\bar{1}$·1991	8	15	23	30	38
81	$\bar{1}$·1943	$\bar{1}$·1895	$\bar{1}$·1847	$\bar{1}$·1797	$\bar{1}$·1747	$\bar{1}$·1697	$\bar{1}$·1646	$\bar{1}$·1594	$\bar{1}$·1542	$\bar{1}$·1489	8	17	25	34	42
82	$\bar{1}$·1436	$\bar{1}$·1381	$\bar{1}$·1326	$\bar{1}$·1271	$\bar{1}$·1214	$\bar{1}$·1157	$\bar{1}$·1099	$\bar{1}$·1040	$\bar{1}$·0981	$\bar{1}$·0920	10	19	29	38	48 *
83	$\bar{1}$·0859	$\bar{1}$·0797	$\bar{1}$·0734	$\bar{1}$·0670	$\bar{1}$·0605	$\bar{1}$·0539	$\bar{1}$·0472	$\bar{1}$·0403	$\bar{1}$·0334	$\bar{1}$·0264	11	22	33	44	55
84	$\bar{1}$·0192	$\bar{1}$·0120	$\bar{1}$·0046	$\bar{2}$·9970	$\bar{2}$·9894	$\bar{2}$·9816	$\bar{2}$·9736	$\bar{2}$·9655	$\bar{2}$·9573	$\bar{2}$·9489	13	26	39	52	66
85	$\bar{2}$·9403	$\bar{2}$·9315	$\bar{2}$·9226	$\bar{2}$·9135	$\bar{2}$·9042	$\bar{2}$·8946	$\bar{2}$·8849	$\bar{2}$·8749	$\bar{2}$·8647	$\bar{2}$·8543	16	32	48	64	80
86	$\bar{2}$·8436	$\bar{2}$·8326	$\bar{2}$·8213	$\bar{2}$·8098	$\bar{2}$·7979	$\bar{2}$·7857	$\bar{2}$·7731	$\bar{2}$·7602	$\bar{2}$·7468	$\bar{2}$·7330	21	41	62	83	104
87	$\bar{2}$·7188	$\bar{2}$·7041	$\bar{2}$·6889	$\bar{2}$·6731	$\bar{2}$·6567	$\bar{2}$·6397	$\bar{2}$·6220	$\bar{2}$·6035	$\bar{2}$·5842	$\bar{2}$·5640					
88	$\bar{2}$·5428	$\bar{2}$·5206	$\bar{2}$·4971	$\bar{2}$·4723	$\bar{2}$·4459	$\bar{2}$·4179	$\bar{2}$·3880	$\bar{2}$·3558	$\bar{2}$·3210	$\bar{2}$·2832					
89	$\bar{2}$·2419	$\bar{2}$·1961	$\bar{2}$·1450	$\bar{2}$·0870	$\bar{2}$·0200	$\bar{3}$·9408	$\bar{3}$·8439	$\bar{3}$·7190	$\bar{3}$·5429	$\bar{3}$·2419					

Table 7. Logarithms of Tangents $\log_{10} \tan x^\circ$

x°	0°·0 0′	0°·1 6′	0°·2 12′	0°·3 18′	0°·4 24′	0°·5 30′	0°·6 36′	0°·7 42′	0°·8 48′	0°·9 54′	1′	2′	3′	4′	5′
0°	$-\infty$	$\bar{3}$·2419	$\bar{3}$·5429	$\bar{3}$·7190	$\bar{3}$·8439	$\bar{3}$·9409	$\bar{2}$·0200	$\bar{2}$·0870	$\bar{2}$·1450	$\bar{2}$·1962					
1	$\bar{2}$·2419	$\bar{2}$·2833	$\bar{2}$·3211	$\bar{2}$·3559	$\bar{2}$·3881	$\bar{2}$·4181	$\bar{2}$·4461	$\bar{2}$·4725	$\bar{2}$·4973	$\bar{2}$·5208					
2	$\bar{2}$·5431	$\bar{2}$·5643	$\bar{2}$·5845	$\bar{2}$·6038	$\bar{2}$·6223	$\bar{2}$·6401	$\bar{2}$·6571	$\bar{2}$·6736	$\bar{2}$·6894	$\bar{2}$·7046	30	60	89	118	147
3	$\bar{2}$·7194	$\bar{2}$·7337	$\bar{2}$·7475	$\bar{2}$·7609	$\bar{2}$·7739	$\bar{2}$·7865	$\bar{2}$·7988	$\bar{2}$·8107	$\bar{2}$·8223	$\bar{2}$·8336	21	42	63	84	105
4	$\bar{2}$·8446	$\bar{2}$·8554	$\bar{2}$·8659	$\bar{2}$·8762	$\bar{2}$·8862	$\bar{2}$·8960	$\bar{2}$·9056	$\bar{2}$·9150	$\bar{2}$·9241	$\bar{2}$·9331	16	33	49	65	81
5	$\bar{2}$·9420	$\bar{2}$·9506	$\bar{2}$·9591	$\bar{2}$·9674	$\bar{2}$·9756	$\bar{2}$·9836	$\bar{2}$·9915	$\bar{2}$·9992	$\bar{1}$·0068	$\bar{1}$·0143	13	27	40	53	66
6	$\bar{1}$·0216	$\bar{1}$·0289	$\bar{1}$·0360	$\bar{1}$·0430	$\bar{1}$·0499	$\bar{1}$·0567	$\bar{1}$·0633	$\bar{1}$·0699	$\bar{1}$·0764	$\bar{1}$·0828	11	23	34	45	56
7	$\bar{1}$·0891	$\bar{1}$·0954	$\bar{1}$·1015	$\bar{1}$·1076	$\bar{1}$·1135	$\bar{1}$·1194	$\bar{1}$·1252	$\bar{1}$·1310	$\bar{1}$·1367	$\bar{1}$·1423	10	20	29	39	49
8	$\bar{1}$·1478	$\bar{1}$·1533	$\bar{1}$·1587	$\bar{1}$·1640	$\bar{1}$·1693	$\bar{1}$·1745	$\bar{1}$·1797	$\bar{1}$·1848	$\bar{1}$·1898	$\bar{1}$·1948	9	17	26	35	43 *
9	$\bar{1}$·1997	$\bar{1}$·2046	$\bar{1}$·2094	$\bar{1}$·2142	$\bar{1}$·2189	$\bar{1}$·2236	$\bar{1}$·2282	$\bar{1}$·2328	$\bar{1}$·2374	$\bar{1}$·2419	8	16	23	31	39
10	$\bar{1}$·2463	$\bar{1}$·2507	$\bar{1}$·2551	$\bar{1}$·2594	$\bar{1}$·2637	$\bar{1}$·2680	$\bar{1}$·2722	$\bar{1}$·2764	$\bar{1}$·2805	$\bar{1}$·2846	7	14	21	28	35
11	$\bar{1}$·2887	$\bar{1}$·2927	$\bar{1}$·2967	$\bar{1}$·3006	$\bar{1}$·3046	$\bar{1}$·3085	$\bar{1}$·3123	$\bar{1}$·3162	$\bar{1}$·3200	$\bar{1}$·3237	6	13	19	26	32
12	$\bar{1}$·3275	$\bar{1}$·3312	$\bar{1}$·3349	$\bar{1}$·3385	$\bar{1}$·3422	$\bar{1}$·3458	$\bar{1}$·3493	$\bar{1}$·3529	$\bar{1}$·3564	$\bar{1}$·3599	6	12	18	24	30
13	$\bar{1}$·3634	$\bar{1}$·3668	$\bar{1}$·3702	$\bar{1}$·3736	$\bar{1}$·3770	$\bar{1}$·3804	$\bar{1}$·3837	$\bar{1}$·3870	$\bar{1}$·3903	$\bar{1}$·3935	6	11	17	22	28
14	$\bar{1}$·3968	$\bar{1}$·4000	$\bar{1}$·4032	$\bar{1}$·4064	$\bar{1}$·4095	$\bar{1}$·4127	$\bar{1}$·4158	$\bar{1}$·4189	$\bar{1}$·4220	$\bar{1}$·4250	5	10	16	21	26
15	$\bar{1}$·4281	$\bar{1}$·4311	$\bar{1}$·4341	$\bar{1}$·4371	$\bar{1}$·4400	$\bar{1}$·4430	$\bar{1}$·4459	$\bar{1}$·4488	$\bar{1}$·4517	$\bar{1}$·4546	5	10	15	20	25
16	$\bar{1}$·4575	$\bar{1}$·4603	$\bar{1}$·4632	$\bar{1}$·4660	$\bar{1}$·4688	$\bar{1}$·4716	$\bar{1}$·4744	$\bar{1}$·4771	$\bar{1}$·4799	$\bar{1}$·4826	5	9	14	19	23
17	$\bar{1}$·4853	$\bar{1}$·4880	$\bar{1}$·4907	$\bar{1}$·4934	$\bar{1}$·4961	$\bar{1}$·4987	$\bar{1}$·5014	$\bar{1}$·5040	$\bar{1}$·5066	$\bar{1}$·5092	4	9	13	18	22
18	$\bar{1}$·5118	$\bar{1}$·5143	$\bar{1}$·5169	$\bar{1}$·5195	$\bar{1}$·5220	$\bar{1}$·5245	$\bar{1}$·5270	$\bar{1}$·5295	$\bar{1}$·5320	$\bar{1}$·5345	4	8	13	17	21
19	$\bar{1}$·5370	$\bar{1}$·5394	$\bar{1}$·5419	$\bar{1}$·5443	$\bar{1}$·5467	$\bar{1}$·5491	$\bar{1}$·5516	$\bar{1}$·5539	$\bar{1}$·5563	$\bar{1}$·5587	4	8	12	16	20
20	$\bar{1}$·5611	$\bar{1}$·5634	$\bar{1}$·5658	$\bar{1}$·5681	$\bar{1}$·5704	$\bar{1}$·5727	$\bar{1}$·5750	$\bar{1}$·5773	$\bar{1}$·5796	$\bar{1}$·5819	4	8	12	15	19
21	$\bar{1}$·5842	$\bar{1}$·5864	$\bar{1}$·5887	$\bar{1}$·5909	$\bar{1}$·5932	$\bar{1}$·5954	$\bar{1}$·5976	$\bar{1}$·5998	$\bar{1}$·6020	$\bar{1}$·6042	4	7	11	15	19
22	$\bar{1}$·6064	$\bar{1}$·6086	$\bar{1}$·6108	$\bar{1}$·6129	$\bar{1}$·6151	$\bar{1}$·6172	$\bar{1}$·6194	$\bar{1}$·6215	$\bar{1}$·6236	$\bar{1}$·6257	4	7	11	14	18
23	$\bar{1}$·6279	$\bar{1}$·6300	$\bar{1}$·6321	$\bar{1}$·6341	$\bar{1}$·6362	$\bar{1}$·6383	$\bar{1}$·6404	$\bar{1}$·6424	$\bar{1}$·6445	$\bar{1}$·6465	3	7	10	14	17
24	$\bar{1}$·6486	$\bar{1}$·6506	$\bar{1}$·6527	$\bar{1}$·6547	$\bar{1}$·6567	$\bar{1}$·6587	$\bar{1}$·6607	$\bar{1}$·6627	$\bar{1}$·6647	$\bar{1}$·6667	3	7	10	13	17
25	$\bar{1}$·6687	$\bar{1}$·6706	$\bar{1}$·6726	$\bar{1}$·6746	$\bar{1}$·6765	$\bar{1}$·6785	$\bar{1}$·6804	$\bar{1}$·6824	$\bar{1}$·6843	$\bar{1}$·6863	3	7	10	13	16
26	$\bar{1}$·6882	$\bar{1}$·6901	$\bar{1}$·6920	$\bar{1}$·6939	$\bar{1}$·6958	$\bar{1}$·6977	$\bar{1}$·6996	$\bar{1}$·7015	$\bar{1}$·7034	$\bar{1}$·7053	3	6	9	13	16
27	$\bar{1}$·7072	$\bar{1}$·7090	$\bar{1}$·7109	$\bar{1}$·7128	$\bar{1}$·7146	$\bar{1}$·7165	$\bar{1}$·7183	$\bar{1}$·7202	$\bar{1}$·7220	$\bar{1}$·7238	3	6	9	12	15
28	$\bar{1}$·7257	$\bar{1}$·7275	$\bar{1}$·7293	$\bar{1}$·7311	$\bar{1}$·7330	$\bar{1}$·7348	$\bar{1}$·7366	$\bar{1}$·7384	$\bar{1}$·7402	$\bar{1}$·7420	3	6	9	12	15
29	$\bar{1}$·7438	$\bar{1}$·7455	$\bar{1}$·7473	$\bar{1}$·7491	$\bar{1}$·7509	$\bar{1}$·7526	$\bar{1}$·7544	$\bar{1}$·7562	$\bar{1}$·7579	$\bar{1}$·7597	3	6	9	12	15
30	$\bar{1}$·7614	$\bar{1}$·7632	$\bar{1}$·7649	$\bar{1}$·7667	$\bar{1}$·7684	$\bar{1}$·7701	$\bar{1}$·7719	$\bar{1}$·7736	$\bar{1}$·7753	$\bar{1}$·7771	3	6	9	12	14
31	$\bar{1}$·7788	$\bar{1}$·7805	$\bar{1}$·7822	$\bar{1}$·7839	$\bar{1}$·7856	$\bar{1}$·7873	$\bar{1}$·7890	$\bar{1}$·7907	$\bar{1}$·7924	$\bar{1}$·7941	3	6	9	11	14
32	$\bar{1}$·7958	$\bar{1}$·7975	$\bar{1}$·7992	$\bar{1}$·8008	$\bar{1}$·8025	$\bar{1}$·8042	$\bar{1}$·8059	$\bar{1}$·8075	$\bar{1}$·8092	$\bar{1}$·8109	3	6	8	11	14
33	$\bar{1}$·8125	$\bar{1}$·8142	$\bar{1}$·8158	$\bar{1}$·8175	$\bar{1}$·8191	$\bar{1}$·8208	$\bar{1}$·8224	$\bar{1}$·8241	$\bar{1}$·8257	$\bar{1}$·8274	3	5	8	11	14
34	$\bar{1}$·8290	$\bar{1}$·8306	$\bar{1}$·8323	$\bar{1}$·8339	$\bar{1}$·8355	$\bar{1}$·8371	$\bar{1}$·8388	$\bar{1}$·8404	$\bar{1}$·8420	$\bar{1}$·8436	3	5	8	11	14
35	$\bar{1}$·8452	$\bar{1}$·8468	$\bar{1}$·8484	$\bar{1}$·8501	$\bar{1}$·8517	$\bar{1}$·8533	$\bar{1}$·8549	$\bar{1}$·8565	$\bar{1}$·8581	$\bar{1}$·8597	3	5	8	11	13
36	$\bar{1}$·8613	$\bar{1}$·8629	$\bar{1}$·8644	$\bar{1}$·8660	$\bar{1}$·8676	$\bar{1}$·8692	$\bar{1}$·8708	$\bar{1}$·8724	$\bar{1}$·8740	$\bar{1}$·8755	3	5	8	11	13
37	$\bar{1}$·8771	$\bar{1}$·8787	$\bar{1}$·8803	$\bar{1}$·8818	$\bar{1}$·8834	$\bar{1}$·8850	$\bar{1}$·8865	$\bar{1}$·8881	$\bar{1}$·8897	$\bar{1}$·8912	3	5	8	10	13
38	$\bar{1}$·8928	$\bar{1}$·8944	$\bar{1}$·8959	$\bar{1}$·8975	$\bar{1}$·8990	$\bar{1}$·9006	$\bar{1}$·9022	$\bar{1}$·9037	$\bar{1}$·9053	$\bar{1}$·9068	3	5	8	10	13
39	$\bar{1}$·9084	$\bar{1}$·9099	$\bar{1}$·9115	$\bar{1}$·9130	$\bar{1}$·9146	$\bar{1}$·9161	$\bar{1}$·9176	$\bar{1}$·9192	$\bar{1}$·9207	$\bar{1}$·9223	3	5	8	10	13
40	$\bar{1}$·9238	$\bar{1}$·9254	$\bar{1}$·9269	$\bar{1}$·9284	$\bar{1}$·9300	$\bar{1}$·9315	$\bar{1}$·9330	$\bar{1}$·9346	$\bar{1}$·9361	$\bar{1}$·9376	3	5	8	10	13
41	$\bar{1}$·9392	$\bar{1}$·9407	$\bar{1}$·9422	$\bar{1}$·9438	$\bar{1}$·9453	$\bar{1}$·9468	$\bar{1}$·9483	$\bar{1}$·9499	$\bar{1}$·9514	$\bar{1}$·9529	3	5	8	10	13
42	$\bar{1}$·9544	$\bar{1}$·9560	$\bar{1}$·9575	$\bar{1}$·9590	$\bar{1}$·9605	$\bar{1}$·9621	$\bar{1}$·9636	$\bar{1}$·9651	$\bar{1}$·9666	$\bar{1}$·9681	3	5	8	10	13
43	$\bar{1}$·9697	$\bar{1}$·9712	$\bar{1}$·9727	$\bar{1}$·9742	$\bar{1}$·9757	$\bar{1}$·9772	$\bar{1}$·9788	$\bar{1}$·9803	$\bar{1}$·9818	$\bar{1}$·9833	3	5	8	10	13
44	$\bar{1}$·9848	$\bar{1}$·9864	$\bar{1}$·9879	$\bar{1}$·9894	$\bar{1}$·9909	$\bar{1}$·9924	$\bar{1}$·9939	$\bar{1}$·9955	$\bar{1}$·9970	$\bar{1}$·9985	3	5	8	10	13

Logarithms of Tangents $\log_{10} \tan x^\circ$

x°	0°·0 0′	0°·1 6′	0°·2 12′	0°·3 18′	0°·4 24′	0°·5 30′	0°·6 36′	0°·7 42′	0°·8 48′	0°·9 54′	1′	2′	3′	4′	5′
45°	0·0000	0·0015	0·0030	0·0045	0·0061	0·0076	0·0091	0·0106	0·0121	0·0136	3	5	8	10	13
46	0·0152	0·0167	0·0182	0·0197	0·0212	0·0228	0·0243	0·0258	0·0273	0·0288	3	5	8	10	13
47	0·0303	0·0319	0·0334	0·0349	0·0364	0·0379	0·0395	0·0410	0·0425	0·0440	3	5	8	10	13
48	0·0456	0·0471	0·0486	0·0501	0·0517	0·0532	0·0547	0·0562	0·0578	0·0593	3	5	8	10	13
49	0·0608	0·0624	0·0639	0·0654	0·0670	0·0685	0·0700	0·0716	0·0731	0·0746	3	5	8	10	13
50	0·0762	0·0777	0·0793	0·0808	0·0824	0·0839	0·0854	0·0870	0·0885	0·0901	3	5	8	10	13
51	0·0916	0·0932	0·0947	0·0963	0·0978	0·0994	0·1010	0·1025	0·1041	0·1056	3	5	8	10	13
52	0·1072	0·1088	0·1103	0·1119	0·1135	0·1150	0·1166	0·1182	0·1197	0·1213	3	5	8	10	13
53	0·1229	0·1245	0·1260	0·1276	0·1292	0·1308	0·1324	0·1340	0·1356	0·1371	3	5	8	11	13
54	0·1387	0·1403	0·1419	0·1435	0·1451	0·1467	0·1483	0·1499	0·1516	0·1532	3	5	8	11	13
55	0·1548	0·1564	0·1580	0·1596	0·1612	0·1629	0·1645	0·1661	0·1677	0·1694	3	5	8	11	14
56	0·1710	0·1726	0·1743	0·1759	0·1776	0·1792	0·1809	0·1825	0·1842	0·1858	3	5	8	11	14
57	0·1875	0·1891	0·1908	0·1925	0·1941	0·1958	0·1975	0·1992	0·2008	0·2025	3	6	8	11	14
58	0·2042	0·2059	0·2076	0·2093	0·2110	0·2127	0·2144	0·2161	0·2178	0·2195	3	6	9	11	14
59	0·2212	0·2229	0·2247	0·2264	0·2281	0·2299	0·2316	0·2333	0·2351	0·2368	3	6	9	12	14
60	0·2386	0·2403	0·2421	0·2438	0·2456	0·2474	0·2491	0·2509	0·2527	0·2545	3	6	9	12	15
61	0·2562	0·2580	0·2598	0·2616	0·2634	0·2652	0·2670	0·2689	0·2707	0·2725	3	6	9	12	15
62	0·2743	0·2762	0·2780	0·2798	0·2817	0·2835	0·2854	0·2872	0·2891	0·2910	3	6	9	12	15
63	0·2928	0·2947	0·2966	0·2985	0·3004	0·3023	0·3042	0·3061	0·3080	0·3099	3	6	9	13	16
64	0·3118	0·3137	0·3157	0·3176	0·3196	0·3215	0·3235	0·3254	0·3274	0·3294	3	6	10	13	16
65	0·3313	0·3333	0·3353	0·3373	0·3393	0·3413	0·3433	0·3453	0·3473	0·3494	3	7	10	13	17
66	0·3514	0·3535	0·3555	0·3576	0·3596	0·3617	0·3638	0·3659	0·3679	0·3700	3	7	10	14	17
67	0·3721	0·3743	0·3764	0·3785	0·3806	0·3828	0·3849	0·3871	0·3892	0·3914	4	7	11	14	18
68	0·3936	0·3958	0·3980	0·4002	0·4024	0·4046	0·4068	0·4091	0·4113	0·4136	4	7	11	15	19
69	0·4158	0·4181	0·4204	0·4227	0·4250	0·4273	0·4296	0·4319	0·4342	0·4366	4	8	12	15	19
70	0·4389	0·4413	0·4437	0·4461	0·4484	0·4509	0·4533	0·4557	0·4581	0·4606	4	8	12	16	20
71	0·4630	0·4655	0·4680	0·4705	0·4730	0·4755	0·4780	0·4805	0·4831	0·4857	4	8	13	17	21
72	0·4882	0·4908	0·4934	0·4960	0·4986	0·5013	0·5039	0·5066	0·5093	0·5120	4	9	13	18	22
73	0·5147	0·5174	0·5201	0·5229	0·5256	0·5284	0·5312	0·5340	0·5368	0·5397	5	9	14	19	23
74	0·5425	0·5454	0·5483	0·5512	0·5541	0·5570	0·5600	0·5629	0·5659	0·5689	5	10	15	20	25
75	0·5719	0·5750	0·5780	0·5811	0·5842	0·5873	0·5905	0·5936	0·5968	0·6000	5	10	16	21	26
76	0·6032	0·6065	0·6097	0·6130	0·6163	0·6196	0·6230	0·6264	0·6298	0·6332	6	11	17	22	28
77	0·6366	0·6401	0·6436	0·6471	0·6507	0·6542	0·6578	0·6615	0·6651	0·6688	6	12	18	24	30
78	0·6725	0·6763	0·6800	0·6838	0·6877	0·6915	0·6954	0·6994	0·7033	0·7073	6	13	19	26	32
79	0·7113	0·7154	0·7195	0·7236	0·7278	0·7320	0·7363	0·7406	0·7449	0·7493	7	14	21	28	35
80	0·7537	0·7581	0·7626	0·7672	0·7718	0·7764	0·7811	0·7858	0·7906	0·7954	8	15	23	31	39
81	0·8003	0·8052	0·8102	0·8152	0·8203	0·8255	0·8307	0·8360	0·8413	0·8467	9	17	26	35	43 *
82	0·8522	0·8577	0·8633	0·8690	0·8748	0·8806	0·8865	0·8924	0·8985	0·9046	10	19	29	39	49
83	0·9109	0·9172	0·9236	0·9301	0·9367	0·9433	0·9501	0·9570	0·9640	0·9711	11	22	34	45	56
84	0·9784	0·9857	0·9932	1·0008	1·0085	1·0164	1·0244	1·0326	1·0409	1·0494	13	26	40	53	66
85	1·0580	1·0669	1·0759	1·0850	1·0944	1·1040	1·1138	1·1238	1·1341	1·1446	16	32	48	65	81
86	1·1554	1·1664	1·1777	1·1893	1·2012	1·2135	1·2261	1·2391	1·2525	1·2663	21	41	62	83	104
87	1·2806	1·2954	1·3106	1·3264	1·3429	1·3599	1·3777	1·3962	1·4155	1·4357					
88	1·4569	1·4792	1·5027	1·5275	1·5539	1·5819	1·6119	1·6441	1·6789	1·7167					
89	1·7581	1·8038	1·8550	1·9130	1·9800	2·0591	2·1561	2·2810	2·4571	2·7581					

Table 8. Squares x^2

x	0	1	2	3	4	5	6	7	8	9	1	2	3	4	5	6	7	8	9
1·0	1·000	1·020	1·040	1·061	1·082	1·103	1·124	1·145	1·166	1·188	2	4	6	8	10	13	15	17	19
1·1	1·210	1·232	1·254	1·277	1·300	1·323	1·346	1·369	1·392	1·416	2	5	7	9	11	14	16	18	21
1·2	1·440	1·464	1·488	1·513	1·538	1·563	1·588	1·613	1·638	1·664	2	5	7	10	12	15	17	20	22
1·3	1·690	1·716	1·742	1·769	1·796	1·823	1·850	1·877	1·904	1·932	3	5	8	11	13	16	19	22	24
1·4	1·960	1·988	2·016	2·045	2·074	2·103	2·132	2·161	2·190	2·220	3	6	9	12	14	17	20	23	26
1·5	2·250	2·280	2·310	2·341	2·372	2·403	2·434	2·465	2·496	2·528	3	6	9	12	15	19	22	25	28
1·6	2·560	2·592	2·624	2·657	2·690	2·723	2·756	2·789	2·822	2·856	3	7	10	13	16	20	23	26	30
1·7	2·890	2·924	2·958	2·993	3·028	3·063	3·098	3·133	3·168	3·204	3	7	10	14	17	21	24	28	31
1·8	3·240	3·276	3·312	3·349	3·386	3·423	3·460	3·497	3·534	3·572	4	7	11	15	18	22	26	30	33
1·9	3·610	3·648	3·686	3·725	3·764	3·803	3·842	3·881	3·920	3·960	4	8	12	16	19	23	27	31	35
2·0	4·000	4·040	4·080	4·121	4·162	4·203	4·244	4·285	4·326	4·368	4	8	12	16	20	25	29	33	37
2·1	4·410	4·452	4·494	4·537	4·580	4·623	4·666	4·709	4·752	4·796	4	9	13	17	21	26	30	34	39
2·2	4·840	4·884	4·928	4·973	5·018	5·063	5·108	5·153	5·198	5·244	4	9	13	18	22	27	31	36	40
2·3	5·290	5·336	5·382	5·429	5·476	5·523	5·570	5·617	5·664	5·712	5	9	14	19	23	28	33	38	42
2·4	5·760	5·808	5·856	5·905	5·954	6·003	6·052	6·101	6·150	6·200	5	10	15	20	24	29	34	39	44
2·5	6·250	6·300	6·350	6·401	6·452	6·503	6·554	6·605	6·656	6·708	5	10	15	20	25	31	36	41	46
2·6	6·760	6·812	6·864	6·917	6·970	7·023	7·076	7·129	7·182	7·236	5	11	16	21	26	32	37	42	48
2·7	7·290	7·344	7·398	7·453	7·508	7·563	7·618	7·673	7·728	7·784	5	11	16	22	27	33	38	44	49
2·8	7·840	7·896	7·952	8·009	8·066	8·123	8·180	8·237	8·294	8·352	6	11	17	23	28	34	40	46	51
2·9	8·410	8·468	8·526	8·585	8·644	8·703	8·762	8·821	8·880	8·940	6	12	18	24	29	35	41	47	53
3·0	9·000	9·060	9·120	9·181	9·242	9·303	9·364	9·425	9·486	9·548	6	12	18	24	30	37	43	49	55
3·1	9·610	9·672	9·734	9·797	9·860	9·923	9·986				6	13	19	25	31	38	44	50	57
3·1								10·05	10·11	10·18	1	1	2	3	3	4	4	5	6
3·2	10·24	10·30	10·37	10·43	10·50	10·56	10·63	10·69	10·76	10·82	1	1	2	3	3	4	5	5	6
3·3	10·89	10·96	11·02	11·09	11·16	11·22	11·29	11·36	11·42	11·49	1	1	2	3	3	4	5	5	6
3·4	11·56	11·63	11·70	11·76	11·83	11·90	11·97	12·04	12·11	12·18	1	1	2	3	3	4	5	6	6
3·5	12·25	12·32	12·39	12·46	12·53	12·60	12·67	12·74	12·82	12·89	1	1	2	3	4	4	5	6	6
3·6	12·96	13·03	13·10	13·18	13·25	13·32	13·40	13·47	13·54	13·62	1	1	2	3	4	4	5	6	7
3·7	13·69	13·76	13·84	13·91	13·99	14·06	14·14	14·21	14·29	14·36	1	2	2	3	4	5	5	6	7
3·8	14·44	14·52	14·59	14·67	14·75	14·82	14·90	14·98	15·05	15·13	1	2	2	3	4	5	5	6	7
3·9	15·21	15·29	15·37	15·44	15·52	15·60	15·68	15·76	15·84	15·92	1	2	2	3	4	5	6	6	7
4·0	16·00	16·08	16·16	16·24	16·32	16·40	16·48	16·56	16·65	16·73	1	2	2	3	4	5	6	6	7
4·1	16·81	16·89	16·97	17·06	17·14	17·22	17·31	17·39	17·47	17·56	1	2	2	3	4	5	6	7	7
4·2	17·64	17·72	17·81	17·89	17·98	18·06	18·15	18·23	18·32	18·40	1	2	3	3	4	5	6	7	8
4·3	18·49	18·58	18·66	18·75	18·84	18·92	19·01	19·10	19·18	19·27	1	2	3	3	4	5	6	7	8
4·4	19·36	19·45	19·54	19·62	19·71	19·80	19·89	19·98	20·07	20·16	1	2	3	4	5	5	6	7	8
4·5	20·25	20·34	20·43	20·52	20·61	20·70	20·79	20·88	20·98	21·07	1	2	3	4	5	5	6	7	8
4·6	21·16	21·25	21·34	21·44	21·53	21·62	21·72	21·81	21·90	22·00	1	2	3	4	5	6	7	7	8
4·7	22·09	22·18	22·28	22·37	22·47	22·56	22·66	22·75	22·85	22·94	1	2	3	4	5	6	7	8	9
4·8	23·04	23·14	23·23	23·33	23·43	23·52	23·62	23·72	23·81	23·91	1	2	3	4	5	6	7	8	9
4·9	24·01	24·11	24·21	24·30	24·40	24·50	24·60	24·70	24·80	24·90	1	2	3	4	5	6	7	8	9
5·0	25·00	25·10	25·20	25·30	25·40	25·50	25·60	25·70	25·81	25·91	1	2	3	4	5	6	7	8	9
5·1	26·01	26·11	26·21	26·32	26·42	26·52	26·63	26·73	26·83	26·94	1	2	3	4	5	6	7	8	9
5·2	27·04	27·14	27·25	27·35	27·46	27·56	27·67	27·77	27·88	27·98	1	2	3	4	5	6	7	8	9
5·3	28·09	28·20	28·30	28·41	28·52	28·62	28·73	28·84	28·94	29·05	1	2	3	4	5	6	7	9	10
5·4	29·16	29·27	29·38	29·48	29·59	29·70	29·81	29·92	30·03	30·14	1	2	3	4	6	7	8	9	10

x	0	1	2	3	4	5	6	7	8	9	1	2	3	4	5	6	7	8	9
5·5	30·25	30·36	30·47	30·58	30·69	30·80	30·91	31·02	31·14	31·25	1	2	3	4	6	7	8	9	10
5·6	31·36	31·47	31·58	31·70	31·81	31·92	32·04	32·15	32·26	32·38	1	2	3	5	6	7	8	9	10
5·7	32·49	32·60	32·72	32·83	32·95	33·06	33·18	33·29	33·41	33·52	1	2	3	5	6	7	8	9	10
5·8	33·64	33·76	33·87	33·99	34·11	34·22	34·34	34·46	34·57	34·69	1	2	4	5	6	7	8	9	11
5·9	34·81	34·93	35·05	35·16	35·28	35·40	35·52	35·64	35·76	35·88	1	2	4	5	6	7	8	10	11
6·0	36·00	36·12	36·24	36·36	36·48	36·60	36·72	36·84	36·97	37·09	1	2	4	5	6	7	9	10	11
6·1	37·21	37·33	37·45	37·58	37·70	37·82	37·95	38·07	38·19	38·32	1	2	4	5	6	7	9	10	11
6·2	38·44	38·56	38·69	38·81	38·94	39·06	39·19	39·31	39·44	39·56	1	3	4	5	6	8	9	10	11
6·3	39·69	39·82	39·94	40·07	40·20	40·32	40·45	40·58	40·70	40·83	1	3	4	5	6	8	9	10	11
6·4	40·96	41·09	41·22	41·34	41·47	41·60	41·73	41·86	41·99	42·12	1	3	4	5	6	8	9	10	12
6·5	42·25	42·38	42·51	42·64	42·77	42·90	43·03	43·16	43·30	43·43	1	3	4	5	7	8	9	10	12
6·6	43·56	43·69	43·82	43·96	44·09	44·22	44·36	44·49	44·62	44·76	1	3	4	5	7	8	9	11	12
6·7	44·89	45·02	45·16	45·29	45·43	45·56	45·70	45·83	45·97	46·10	1	3	4	5	7	8	9	11	12
6·8	46·24	46·38	46·51	46·65	46·79	46·92	47·06	47·20	47·33	47·47	1	3	4	5	7	8	10	11	12
6·9	47·61	47·75	47·89	48·02	48·16	48·30	48·44	48·58	48·72	48·86	1	3	4	6	7	8	10	11	13
7·0	49·00	49·14	49·28	49·42	49·56	49·70	49·84	49·98	50·13	50·27	1	3	4	6	7	8	10	11	13
7·1	50·41	50·55	50·69	50·84	50·98	51·12	51·27	51·41	51·55	51·70	1	3	4	6	7	9	10	11	13
7·2	51·84	51·98	52·13	52·27	52·42	52·56	52·71	52·85	53·00	53·14	1	3	4	6	7	9	10	12	13
7·3	53·29	53·44	53·58	53·73	53·88	54·02	54·17	54·32	54·46	54·61	1	3	4	6	7	9	10	12	13
7·4	54·76	54·91	55·06	55·20	55·35	55·50	55·65	55·80	55·95	56·10	1	3	4	6	7	9	10	12	13
7·5	56·25	56·40	56·55	56·70	56·85	57·00	57·15	57·30	57·46	57·61	2	3	5	6	8	9	11	12	14
7·6	57·76	57·91	58·06	58·22	58·37	58·52	58·68	58·83	58·98	59·14	2	3	5	6	8	9	11	12	14
7·7	59·29	59·44	59·60	59·75	59·91	60·06	60·22	60·37	60·53	60·68	2	3	5	6	8	9	11	12	14
7·8	60·84	61·00	61·15	61·31	61·47	61·62	61·78	61·94	62·09	62·25	2	3	5	6	8	9	11	13	14
7·9	62·41	62·57	62·73	62·88	63·04	63·20	63·36	63·52	63·68	63·84	2	3	5	6	8	10	11	13	14
8·0	64·00	64·16	64·32	64·48	64·64	64·80	64·96	65·12	65·29	65·45	2	3	5	6	8	10	11	13	14
8·1	65·61	65·77	65·93	66·10	66·26	66·42	66·59	66·75	66·91	67·08	2	3	5	7	8	10	11	13	15
8·2	67·24	67·40	67·57	67·73	67·90	68·06	68·23	68·39	68·56	68·72	2	3	5	7	8	10	12	13	15
8·3	68·89	69·06	69·22	69·39	69·56	69·72	69·89	70·06	70·22	70·39	2	3	5	7	8	10	12	13	15
8·4	70·56	70·73	70·90	71·06	71·23	71·40	71·57	71·74	71·91	72·08	2	3	5	7	8	10	12	14	15
8·5	72·25	72·42	72·59	72·76	72·93	73·10	73·27	73·44	73·62	73·79	2	3	5	7	9	10	12	14	15
8·6	73·96	74·13	74·30	74·48	74·65	74·82	75·00	75·17	75·34	75·52	2	3	5	7	9	10	12	14	16
8·7	75·69	75·86	76·04	76·21	76·39	76·56	76·74	76·91	77·09	77·26	2	4	5	7	9	11	12	14	16
8·8	77·44	77·62	77·79	77·97	78·15	78·32	78·50	78·68	78·85	79·03	2	4	5	7	9	11	12	14	16
8·9	79·21	79·39	79·57	79·74	79·92	80·10	80·28	80·46	80·64	80·82	2	4	5	7	9	11	13	14	16
9·0	81·00	81·18	81·36	81·54	81·72	81·90	82·08	82·26	82·45	82·63	2	4	5	7	9	11	13	14	16
9·1	82·81	82·99	83·17	83·36	83·54	83·72	83·91	84·09	84·27	84·46	2	4	5	7	9	11	13	15	16
9·2	84·64	84·82	85·01	85·19	85·38	85·56	85·75	85·93	86·12	86·30	2	4	6	7	9	11	13	15	17
9·3	86·49	86·68	86·86	87·05	87·24	87·42	87·61	87·80	87·98	88·17	2	4	6	7	9	11	13	15	17
9·4	88·36	88·55	88·74	88·92	89·11	89·30	89·49	89·68	89·87	90·06	2	4	6	8	9	11	13	15	17
9·5	90·25	90·44	90·63	90·82	91·01	91·20	91·39	91·58	91·78	91·97	2	4	6	8	10	11	13	15	17
9·6	92·16	92·35	92·54	92·74	92·93	93·12	93·32	93·51	93·70	93·90	2	4	6	8	10	12	14	15	17
9·7	94·09	94·28	94·48	94·67	94·87	95·06	95·26	95·45	95·65	95·84	2	4	6	8	10	12	14	16	18
9·8	96·04	96·24	96·43	96·63	96·83	97·02	97·22	97·42	97·61	97·81	2	4	6	8	10	12	14	16	18
9·9	98·01	98·21	98·41	98·60	98·80	99·00	99·20	99·40	99·60	99·80	2	4	6	8	10	12	14	16	18

Table 9. Square roots of numbers 1 to 10 $\sqrt{x}$

x	0	1	2	3	4	5	6	7	8	9	1	2	3	4	5	6	7	8	9
1·0	1·0000	1·0050	1·0100	1·0149	1·0198	1·0247	1·0296	1·0344	1·0392	1·0440	5	10	15	20	24	29	34	39	44
1·1	1·0488	1·0536	1·0583	1·0630	1·0677	1·0724	1·0770	1·0817	1·0863	1·0909	5	9	14	19	23	28	33	37	42
1·2	1·0954	1·1000	1·1045	1·1091	1·1136	1·1180	1·1225	1·1269	1·1314	1·1358	4	9	13	18	22	27	31	36	40
1·3	1·1402	1·1446	1·1489	1·1533	1·1576	1·1619	1·1662	1·1705	1·1747	1·1790	4	9	13	17	22	26	30	35	39
1·4	1·1832	1·1874	1·1916	1·1958	1·2000	1·2042	1·2083	1·2124	1·2166	1·2207	4	8	12	17	21	25	29	33	37
1·5	1·2247	1·2288	1·2329	1·2369	1·2410	1·2450	1·2490	1·2530	1·2570	1·2610	4	8	12	16	20	24	28	32	36
1·6	1·2649	1·2689	1·2728	1·2767	1·2806	1·2845	1·2884	1·2923	1·2961	1·3000	4	8	12	16	20	23	27	31	35
1·7	1·3038	1·3077	1·3115	1·3153	1·3191	1·3229	1·3266	1·3304	1·3342	1·3379	4	8	11	15	19	23	27	30	34
1·8	1·3416	1·3454	1·3491	1·3528	1·3565	1·3601	1·3638	1·3675	1·3711	1·3748	4	7	11	15	18	22	26	29	33
1·9	1·3784	1·3820	1·3856	1·3892	1·3928	1·3964	1·4000	1·4036	1·4071	1·4107	4	7	11	14	18	22	25	29	32
2·0	1·4142	1·4177	1·4213	1·4248	1·4283	1·4318	1·4353	1·4387	1·4422	1·4457	4	7	11	14	17	21	24	28	31
2·1	1·4491	1·4526	1·4560	1·4595	1·4629	1·4663	1·4697	1·4731	1·4765	1·4799	3	7	10	14	17	21	24	27	31
2·2	1·4832	1·4866	1·4900	1·4933	1·4967	1·5000	1·5033	1·5067	1·5100	1·5133	3	7	10	13	17	20	23	27	30
2·3	1·5166	1·5199	1·5232	1·5264	1·5297	1·5330	1·5362	1·5395	1·5427	1·5460	3	7	10	13	16	20	23	26	29
2·4	1·5492	1·5524	1·5556	1·5588	1·5620	1·5652	1·5684	1·5716	1·5748	1·5780	3	6	10	13	16	19	22	26	29
2·5	1·5811	1·5843	1·5875	1·5906	1·5937	1·5969	1·6000	1·6031	1·6062	1·6093	3	6	9	13	16	19	22	25	28
2·6	1·6125	1·6155	1·6186	1·6217	1·6248	1·6279	1·6310	1·6340	1·6371	1·6401	3	6	9	12	15	18	22	25	28
2·7	1·6432	1·6462	1·6492	1·6523	1·6553	1·6583	1·6613	1·6643	1·6673	1·6703	3	6	9	12	15	18	21	24	27
2·8	1·6733	1·6763	1·6793	1·6823	1·6852	1·6882	1·6912	1·6941	1·6971	1·7000	3	6	9	12	15	18	21	24	27
2·9	1·7029	1·7059	1·7088	1·7117	1·7146	1·7176	1·7205	1·7234	1·7263	1·7292	3	6	9	12	15	17	20	23	26
3·0	1·7321	1·7349	1·7378	1·7407	1·7436	1·7464	1·7493	1·7521	1·7550	1·7578	3	6	9	11	14	17	20	23	26
3·1	1·7607	1·7635	1·7664	1·7692	1·7720	1·7748	1·7776	1·7804	1·7833	1·7861	3	6	8	11	14	17	20	23	25
3·2	1·7889	1·7916	1·7944	1·7972	1·8000	1·8028	1·8055	1·8083	1·8111	1·8138	3	6	8	11	14	17	19	22	25
3·3	1·8166	1·8193	1·8221	1·8248	1·8276	1·8303	1·8330	1·8358	1·8385	1·8412	3	5	8	11	14	16	19	22	25
3·4	1·8439	1·8466	1·8493	1·8520	1·8547	1·8574	1·8601	1·8628	1·8655	1·8682	3	5	8	11	13	16	19	22	24
3·5	1·8708	1·8735	1·8762	1·8788	1·8815	1·8841	1·8868	1·8894	1·8921	1·8947	3	5	8	11	13	16	19	21	24
3·6	1·8974	1·9000	1·9026	1·9053	1·9079	1·9105	1·9131	1·9157	1·9183	1·9209	3	5	8	10	13	16	18	21	24
3·7	1·9235	1·9261	1·9287	1·9313	1·9339	1·9365	1·9391	1·9416	1·9442	1·9468	3	5	8	10	13	16	18	21	23
3·8	1·9494	1·9519	1·9545	1·9570	1·9596	1·9621	1·9647	1·9672	1·9698	1·9723	3	5	8	10	13	15	18	20	23
3·9	1·9748	1·9774	1·9799	1·9824	1·9849	1·9875	1·9900	1·9925	1·9950	1·9975	3	5	8	10	13	15	18	20	23
4·0	2·0000	2·0025	2·0050	2·0075	2·0100	2·0125	2·0149	2·0174	2·0199	2·0224	2	5	7	10	12	15	17	20	22
4·1	2·0248	2·0273	2·0298	2·0322	2·0347	2·0372	2·0396	2·0421	2·0445	2·0469	2	5	7	10	12	15	17	20	22
4·2	2·0494	2·0518	2·0543	2·0567	2·0591	2·0616	2·0640	2·0664	2·0688	2·0712	2	5	7	10	12	15	17	19	22
4·3	2·0736	2·0761	2·0785	2·0809	2·0833	2·0857	2·0881	2·0905	2·0928	2·0952	2	5	7	10	12	14	17	19	22
4·4	2·0976	2·1000	2·1024	2·1048	2·1071	2·1095	2·1119	2·1142	2·1166	2·1190	2	5	7	9	12	14	17	19	21
4·5	2·1213	2·1237	2·1260	2·1284	2·1307	2·1331	2·1354	2·1378	2·1401	2·1424	2	5	7	9	12	14	16	19	21
4·6	2·1448	2·1471	2·1494	2·1517	2·1541	2·1564	2·1587	2·1610	2·1633	2·1656	2	5	7	9	12	14	16	19	21
4·7	2·1679	2·1703	2·1726	2·1749	2·1772	2·1794	2·1817	2·1840	2·1863	2·1886	2	5	7	9	11	14	16	18	21
4·8	2·1909	2·1932	2·1954	2·1977	2·2000	2·2023	2·2045	2·2068	2·2091	2·2113	2	5	7	9	11	14	16	18	20
4·9	2·2136	2·2159	2·2181	2·2204	2·2226	2·2249	2·2271	2·2293	2·2316	2·2338	2	4	7	9	11	13	16	18	20
5·0	2·2361	2·2383	2·2405	2·2428	2·2450	2·2472	2·2494	2·2517	2·2539	2·2561	2	4	7	9	11	13	16	18	20
5·1	2·2583	2·2605	2·2627	2·2650	2·2672	2·2694	2·2716	2·2738	2·2760	2·2782	2	4	7	9	11	13	15	18	20
5·2	2·2804	2·2825	2·2847	2·2869	2·2891	2·2913	2·2935	2·2956	2·2978	2·3000	2	4	7	9	11	13	15	17	20
5·3	2·3022	2·3043	2·3065	2·3087	2·3108	2·3130	2·3152	2·3173	2·3195	2·3216	2	4	6	9	11	13	15	17	19
5·4	2·3238	2·3259	2·3281	2·3302	2·3324	2·3345	2·3367	2·3388	2·3409	2·3431	2	4	6	9	11	13	15	17	19

Square roots of numbers 1 to 10

$\sqrt{x}$

x	0	1	2	3	4	5	6	7	8	9	1	2	3	4	5	6	7	8	9
5·5	2·3452	2·3473	2·3495	2·3516	2·3537	2·3558	2·3580	2·3601	2·3622	2·3643	2	4	6	9	11	13	15	17	19
5·6	2·3664	2·3685	2·3707	2·3728	2·3749	2·3770	2·3791	2·3812	2·3833	2·3854	2	4	6	8	11	13	15	17	19
5·7	2·3875	2·3896	2·3917	2·3937	2·3958	2·3979	2·4000	2·4021	2·4042	2·4062	2	4	6	8	10	13	15	17	19
5·8	2·4083	2·4104	2·4125	2·4145	2·4166	2·4187	2·4207	2·4228	2·4249	2·4269	2	4	6	8	10	12	14	17	19
5·9	2·4290	2·4310	2·4331	2·4352	2·4372	2·4393	2·4413	2·4434	2·4454	2·4474	2	4	6	8	10	12	14	16	18
6·0	2·4495	2·4515	2·4536	2·4556	2·4576	2·4597	2·4617	2·4637	2·4658	2·4678	2	4	6	8	10	12	14	16	18
6·1	2·4698	2·4718	2·4739	2·4759	2·4779	2·4799	2·4819	2·4839	2·4860	2·4880	2	4	6	8	10	12	14	16	18
6·2	2·4900	2·4920	2·4940	2·4960	2·4980	2·5000	2·5020	2·5040	2·5060	2·5080	2	4	6	8	10	12	14	16	18
6·3	2·5100	2·5120	2·5140	2·5159	2·5179	2·5199	2·5219	2·5239	2·5259	2·5278	2	4	6	8	10	12	14	16	18
6·4	2·5298	2·5318	2·5338	2·5357	2·5377	2·5397	2·5417	2·5436	2·5456	2·5475	2	4	6	8	10	12	14	16	18
6·5	2·5495	2·5515	2·5534	2·5554	2·5573	2·5593	2·5612	2·5632	2·5652	2·5671	2	4	6	8	10	12	14	16	18
6·6	2·5690	2·5710	2·5729	2·5749	2·5768	2·5788	2·5807	2·5826	2·5846	2·5865	2	4	6	8	10	12	14	16	17
6·7	2·5884	2·5904	2·5923	2·5942	2·5962	2·5981	2·6000	2·6019	2·6038	2·6058	2	4	6	8	10	12	13	15	17
6·8	2·6077	2·6096	2·6115	2·6134	2·6153	2·6173	2·6192	2·6211	2·6230	2·6249	2	4	6	8	10	11	13	15	17
6·9	2·6268	2·6287	2·6306	2·6325	2·6344	2·6363	2·6382	2·6401	2·6420	2·6439	2	4	6	8	9	11	13	15	17
7·0	2·6458	2·6476	2·6495	2·6514	2·6533	2·6552	2·6571	2·6589	2·6608	2·6627	2	4	6	8	9	11	13	15	17
7·1	2·6646	2·6665	2·6683	2·6702	2·6721	2·6739	2·6758	2·6777	2·6796	2·6814	2	4	6	7	9	11	13	15	17
7·2	2·6833	2·6851	2·6870	2·6889	2·6907	2·6926	2·6944	2·6963	2·6981	2·7000	2	4	6	7	9	11	13	15	17
7·3	2·7019	2·7037	2·7055	2·7074	2·7092	2·7111	2·7129	2·7148	2·7166	2·7185	2	4	6	7	9	11	13	15	17
7·4	2·7203	2·7221	2·7240	2·7258	2·7276	2·7295	2·7313	2·7331	2·7350	2·7368	2	4	6	7	9	11	13	15	16
7·5	2·7386	2·7404	2·7423	2·7441	2·7459	2·7477	2·7495	2·7514	2·7532	2·7550	2	4	5	7	9	11	13	15	16
7·6	2·7568	2·7586	2·7604	2·7622	2·7641	2·7659	2·7677	2·7695	2·7713	2·7731	2	4	5	7	9	11	13	14	16
7·7	2·7749	2·7767	2·7785	2·7803	2·7821	2·7839	2·7857	2·7875	2·7893	2·7911	2	4	5	7	9	11	13	14	16
7·8	2·7928	2·7946	2·7964	2·7982	2·8000	2·8018	2·8036	2·8054	2·8071	2·8089	2	4	5	7	9	11	12	14	16
7·9	2·8107	2·8125	2·8142	2·8160	2·8178	2·8196	2·8213	2·8231	2·8249	2·8267	2	4	5	7	9	11	12	14	16
8·0	2·8284	2·8302	2·8320	2·8337	2·8355	2·8373	2·8390	2·8408	2·8425	2·8443	2	4	5	7	9	11	12	14	16
8·1	2·8460	2·8478	2·8496	2·8513	2·8531	2·8548	2·8566	2·8583	2·8601	2·8618	2	4	5	7	9	11	12	14	16
8·2	2·8636	2·8653	2·8671	2·8688	2·8705	2·8723	2·8740	2·8758	2·8775	2·8792	2	3	5	7	9	10	12	14	16
8·3	2·8810	2·8827	2·8844	2·8862	2·8879	2·8896	2·8914	2·8931	2·8948	2·8965	2	3	5	7	9	10	12	14	16
8·4	2·8983	2·9000	2·9017	2·9034	2·9052	2·9069	2·9086	2·9103	2·9120	2·9138	2	3	5	7	9	10	12	14	15
8·5	2·9155	2·9172	2·9189	2·9206	2·9223	2·9240	2·9257	2·9275	2·9292	2·9309	2	3	5	7	9	10	12	14	15
8·6	2·9326	2·9343	2·9360	2·9377	2·9394	2·9411	2·9428	2·9445	2·9462	2·9479	2	3	5	7	9	10	12	14	15
8·7	2·9496	2·9513	2·9530	2·9547	2·9563	2·9580	2·9597	2·9614	2·9631	2·9648	2	3	5	7	8	10	12	14	15
8·8	2·9665	2·9682	2·9698	2·9715	2·9732	2·9749	2·9766	2·9783	2·9799	2·9816	2	3	5	7	8	10	12	13	15
8·9	2·9833	2·9850	2·9866	2·9883	2·9900	2·9917	2·9933	2·9950	2·9967	2·9983	2	3	5	7	8	10	12	13	15
9·0	3·0000	3·0017	3·0033	3·0050	3·0067	3·0083	3·0100	3·0116	3·0133	3·0150	2	3	5	7	8	10	12	13	15
9·1	3·0166	3·0183	3·0199	3·0216	3·0232	3·0249	3·0265	3·0282	3·0299	3·0315	2	3	5	7	8	10	12	13	15
9·2	3·0332	3·0348	3·0364	3·0381	3·0397	3·0414	3·0430	3·0447	3·0463	3·0480	2	3	5	7	8	10	12	13	15
9·3	3·0496	3·0512	3·0529	3·0545	3·0561	3·0578	3·0594	3·0610	3·0627	3·0643	2	3	5	7	8	10	11	13	15
9·4	3·0659	3·0676	3·0692	3·0708	3·0725	3·0741	3·0757	3·0773	3·0790	3·0806	2	3	5	7	8	10	11	13	15
9·5	3·0822	3·0838	3·0854	3·0871	3·0887	3·0903	3·0919	3·0935	3·0952	3·0968	2	3	5	6	8	10	11	13	15
9·6	3·0984	3·1000	3·1016	3·1032	3·1048	3·1064	3·1081	3·1097	3·1113	3·1129	2	3	5	6	8	10	11	13	14
9·7	3·1145	3·1161	3·1177	3·1193	3·1209	3·1225	3·1241	3·1257	3·1273	3·1289	2	3	5	6	8	10	11	13	14
9·8	3·1305	3·1321	3·1337	3·1353	3·1369	3·1385	3·1401	3·1417	3·1432	3·1448	2	3	5	6	8	10	11	13	14
9·9	3·1464	3·1480	3·1496	3·1512	3·1528	3·1544	3·1559	3·1575	3·1591	3·1607	2	3	5	6	8	10	11	13	14

Table 10. Square roots of numbers 10 to 100 $\sqrt{x}$

x	0	1	2	3	4	5	6	7	8	9	1	2	3	4	5	6	7	8	9
10	3·1623	3·1780	3·1937	3·2094	3·2249	3·2404	3·2558	3·2711	3·2863	3·3015	16	31	46	62	77	93	108	124	139
11	3·3166	3·3317	3·3466	3·3615	3·3764	3·3912	3·4059	3·4205	3·4351	3·4496	15	30	44	59	74	89	104	118	133
12	3·4641	3·4785	3·4928	3·5071	3·5214	3·5355	3·5496	3·5637	3·5777	3·5917	14	28	43	57	71	85	99	113	128
13	3·6056	3·6194	3·6332	3·6469	3·6606	3·6742	3·6878	3·7014	3·7148	3·7283	14	27	41	55	68	82	95	109	123
14	3·7417	3·7550	3·7683	3·7815	3·7947	3·8079	3·8210	3·8341	3·8471	3·8601	13	26	40	53	66	79	92	105	118
15	3·8730	3·8859	3·8987	3·9115	3·9243	3·9370	3·9497	3·9623	3·9749	3·9875	13	25	38	51	64	76	89	102	115
16	4·0000	4·0125	4·0249	4·0373	4·0497	4·0620	4·0743	4·0866	4·0988	4·1110	12	25	37	49	62	74	86	99	111
17	4·1231	4·1352	4·1473	4·1593	4·1713	4·1833	4·1952	4·2071	4·2190	4·2308	12	24	36	48	60	72	84	96	108
18	4·2426	4·2544	4·2661	4·2778	4·2895	4·3012	4·3128	4·3243	4·3359	4·3474	12	23	35	47	58	70	82	93	105
19	4·3589	4·3704	4·3818	4·3932	4·4045	4·4159	4·4272	4·4385	4·4497	4·4609	11	23	34	45	57	68	79	91	102
20	4·4721	4·4833	4·4944	4·5056	4·5166	4·5277	4·5387	4·5497	4·5607	4·5717	11	22	33	44	55	66	77	88	100
21	4·5826	4·5935	4·6043	4·6152	4·6260	4·6368	4·6476	4·6583	4·6690	4·6797	11	22	32	43	54	65	76	86	97
22	4·6904	4·7011	4·7117	4·7223	4·7329	4·7434	4·7539	4·7645	4·7749	4·7854	11	21	32	42	53	63	74	84	95
23	4·7958	4·8062	4·8166	4·8270	4·8374	4·8477	4·8580	4·8683	4·8785	4·8888	10	21	31	41	52	62	72	83	93
24	4·8990	4·9092	4·9193	4·9295	4·9396	4·9497	4·9598	4·9699	4·9800	4·9900	10	20	30	40	51	61	71	81	91
25	5·0000	5·0100	5·0200	5·0299	5·0398	5·0498	5·0596	5·0695	5·0794	5·0892	10	20	30	40	50	59	69	79	89
26	5·0990	5·1088	5·1186	5·1284	5·1381	5·1478	5·1575	5·1672	5·1769	5·1865	10	19	29	39	49	58	68	78	88
27	5·1962	5·2058	5·2154	5·2249	5·2345	5·2440	5·2536	5·2631	5·2726	5·2820	10	19	29	38	48	57	67	76	86
28	5·2915	5·3009	5·3104	5·3198	5·3292	5·3385	5·3479	5·3572	5·3666	5·3759	9	19	28	38	47	56	66	75	84
29	5·3852	5·3944	5·4037	5·4129	5·4222	5·4314	5·4406	5·4498	5·4589	5·4681	9	18	28	37	46	55	65	74	83
30	5·4772	5·4863	5·4955	5·5045	5·5136	5·5227	5·5317	5·5408	5·5498	5·5588	9	18	27	36	45	54	63	73	82
31	5·5678	5·5767	5·5857	5·5946	5·6036	5·6125	5·6214	5·6303	5·6391	5·6480	9	18	27	36	45	54	62	71	80
32	5·6569	5·6657	5·6745	5·6833	5·6921	5·7009	5·7096	5·7184	5·7271	5·7359	9	18	26	35	44	53	61	70	79
33	5·7446	5·7533	5·7619	5·7706	5·7793	5·7879	5·7966	5·8052	5·8138	5·8224	9	17	26	35	43	52	61	69	78
34	5·8310	5·8395	5·8481	5·8566	5·8652	5·8737	5·8822	5·8907	5·8992	5·9076	9	17	26	34	43	51	60	68	77
35	5·9161	5·9245	5·9330	5·9414	5·9498	5·9582	5·9666	5·9749	5·9833	5·9917	8	17	25	34	42	50	59	67	76
36	6·0000	6·0083	6·0166	6·0249	6·0332	6·0415	6·0498	6·0581	6·0663	6·0745	8	17	25	33	41	50	58	66	75
37	6·0828	6·0910	6·0992	6·1074	6·1156	6·1237	6·1319	6·1400	6·1482	6·1563	8	16	25	33	41	49	57	65	74
38	6·1644	6·1725	6·1806	6·1887	6·1968	6·2048	6·2129	6·2209	6·2290	6·2370	8	16	24	32	40	48	56	65	73
39	6·2450	6·2530	6·2610	6·2690	6·2769	6·2849	6·2929	6·3008	6·3087	6·3166	8	16	24	32	40	48	56	64	72
40	6·3246	6·3325	6·3403	6·3482	6·3561	6·3640	6·3718	6·3797	6·3875	6·3953	8	16	24	31	39	47	55	63	71
41	6·4031	6·4109	6·4187	6·4265	6·4343	6·4420	6·4498	6·4576	6·4653	6·4730	8	16	23	31	39	47	54	62	70
42	6·4807	6·4885	6·4962	6·5038	6·5115	6·5192	6·5269	6·5345	6·5422	6·5498	8	15	23	31	38	46	54	61	69
43	6·5574	6·5651	6·5727	6·5803	6·5879	6·5955	6·6030	6·6106	6·6182	6·6257	8	15	23	30	38	46	53	61	68
44	6·6332	6·6408	6·6483	6·6558	6·6633	6·6708	6·6783	6·6858	6·6933	6·7007	8	15	23	30	38	45	53	60	68
45	6·7082	6·7157	6·7231	6·7305	6·7380	6·7454	6·7528	6·7602	6·7676	6·7750	7	15	22	30	37	45	52	59	67
46	6·7823	6·7897	6·7971	6·8044	6·8118	6·8191	6·8264	6·8337	6·8411	6·8484	7	15	22	29	37	44	51	59	66
47	6·8557	6·8629	6·8702	6·8775	6·8848	6·8920	6·8993	6·9065	6·9138	6·9210	7	15	22	29	36	44	51	58	65
48	6·9282	6·9354	6·9426	6·9498	6·9570	6·9642	6·9714	6·9785	6·9857	6·9929	7	14	22	29	36	43	50	57	65
49	7·0000	7·0071	7·0143	7·0214	7·0285	7·0356	7·0427	7·0498	7·0569	7·0640	7	14	21	28	36	43	50	57	64
50	7·0711	7·0781	7·0852	7·0922	7·0993	7·1063	7·1134	7·1204	7·1274	7·1344	7	14	21	28	35	42	49	56	63
51	7·1414	7·1484	7·1554	7·1624	7·1694	7·1764	7·1833	7·1903	7·1972	7·2042	7	14	21	28	35	42	49	56	63
52	7·2111	7·2180	7·2250	7·2319	7·2388	7·2457	7·2526	7·2595	7·2664	7·2732	7	14	21	28	35	41	48	55	62
53	7·2801	7·2870	7·2938	7·3007	7·3075	7·3144	7·3212	7·3280	7·3348	7·3417	7	14	21	27	34	41	48	55	62
54	7·3485	7·3553	7·3621	7·3689	7·3756	7·3824	7·3892	7·3959	7·4027	7·4095	7	14	20	27	34	41	47	54	61

x	0	1	2	3	4	5	6	7	8	9	1	2	3	4	5	6	7	8	9
55	7·4162	7·4229	7·4297	7·4364	7·4431	7·4498	7·4565	7·4632	7·4699	7·4766	7	13	20	27	34	40	47	54	60
56	7·4833	7·4900	7·4967	7·5033	7·5100	7·5166	7·5233	7·5299	7·5366	7·5432	7	13	20	27	33	40	47	53	60
57	7·5498	7·5565	7·5631	7·5697	7·5763	7·5829	7·5895	7·5961	7·6026	7·6092	7	13	20	26	33	40	46	53	59
58	7·6158	7·6223	7·6289	7·6354	7·6420	7·6485	7·6551	7·6616	7·6681	7·6746	7	13	20	26	33	39	46	52	59
59	7·6811	7·6877	7·6942	7·7006	7·7071	7·7136	7·7201	7·7266	7·7330	7·7395	6	13	19	26	32	39	45	52	58
60	7·7460	7·7524	7·7589	7·7653	7·7717	7·7782	7·7846	7·7910	7·7974	7·8038	6	13	19	26	32	39	45	51	58
61	7·8102	7·8166	7·8230	7·8294	7·8358	7·8422	7·8486	7·8549	7·8613	7·8677	6	13	19	26	32	38	45	51	57
62	7·8740	7·8804	7·8867	7·8930	7·8994	7·9057	7·9120	7·9183	7·9246	7·9310	6	13	19	25	32	38	44	51	57
63	7·9373	7·9436	7·9498	7·9561	7·9624	7·9687	7·9750	7·9812	7·9875	7·9937	6	13	19	25	31	38	44	50	57
64	8·0000	8·0062	8·0125	8·0187	8·0250	8·0312	8·0374	8·0436	8·0498	8·0561	6	12	19	25	31	37	44	50	56
65	8·0623	8·0685	8·0747	8·0808	8·0870	8·0932	8·0994	8·1056	8·1117	8·1179	6	12	19	25	31	37	43	49	56
66	8·1240	8·1302	8·1363	8·1425	8·1486	8·1548	8·1609	8·1670	8·1731	8·1792	6	12	18	25	31	37	43	49	55
67	8·1854	8·1915	8·1976	8·2037	8·2098	8·2158	8·2219	8·2280	8·2341	8·2401	6	12	18	24	30	37	43	49	55
68	8·2462	8·2523	8·2583	8·2644	8·2704	8·2765	8·2825	8·2885	8·2946	8·3006	6	12	18	24	30	36	42	48	54
69	8·3066	8·3126	8·3187	8·3247	8·3307	8·3367	8·3427	8·3487	8·3546	8·3606	6	12	18	24	30	36	42	48	54
70	8·3666	8·3726	8·3785	8·3845	8·3905	8·3964	8·4024	8·4083	8·4143	8·4202	6	12	18	24	30	36	42	48	54
71	8·4261	8·4321	8·4380	8·4439	8·4499	8·4558	8·4617	8·4676	8·4735	8·4794	6	12	18	24	30	35	41	47	53
72	8·4853	8·4912	8·4971	8·5029	8·5088	8·5147	8·5206	8·5264	8·5323	8·5381	6	12	18	24	29	35	41	47	53
73	8·5440	8·5499	8·5557	8·5615	8·5674	8·5732	8·5790	8·5849	8·5907	8·5965	6	12	18	23	29	35	41	47	53
74	8·6023	8·6081	8·6139	8·6197	8·6255	8·6313	8·6371	8·6429	8·6487	8·6545	6	12	17	23	29	35	41	46	52
75	8·6603	8·6660	8·6718	8·6776	8·6833	8·6891	8·6948	8·7006	8·7063	8·7121	6	12	17	23	29	35	40	46	52
76	8·7178	8·7235	8·7293	8·7350	8·7407	8·7464	8·7521	8·7579	8·7636	8·7693	6	11	17	23	29	34	40	46	51
77	8·7750	8·7807	8·7864	8·7920	8·7977	8·8034	8·8091	8·8148	8·8204	8·8261	6	11	17	23	28	34	40	45	51
78	8·8318	8·8374	8·8431	8·8487	8·8544	8·8600	8·8657	8·8713	8·8769	8·8826	6	11	17	23	28	34	40	45	51
79	8·8882	8·8938	8·8994	8·9051	8·9107	8·9163	8·9219	8·9275	8·9331	8·9387	6	11	17	22	28	34	39	45	50
80	8·9443	8·9499	8·9554	8·9610	8·9666	8·9722	8·9778	8·9833	8·9889	8·9944	6	11	17	22	28	33	39	45	50
81	9·0000	9·0056	9·0111	9·0167	9·0222	9·0277	9·0333	9·0388	9·0443	9·0499	6	11	17	22	28	33	39	44	50
82	9·0554	9·0609	9·0664	9·0719	9·0774	9·0830	9·0885	9·0940	9·0995	9·1049	6	11	17	22	28	33	39	44	50
83	9·1104	9·1159	9·1214	9·1269	9·1324	9·1378	9·1433	9·1488	9·1542	9·1597	5	11	16	22	27	33	38	44	49
84	9·1652	9·1706	9·1761	9·1815	9·1869	9·1924	9·1978	9·2033	9·2087	9·2141	5	11	16	22	27	33	38	44	49
85	9·2195	9·2250	9·2304	9·2358	9·2412	9·2466	9·2520	9·2574	9·2628	9·2682	5	11	16	22	27	32	38	43	49
86	9·2736	9·2790	9·2844	9·2898	9·2952	9·3005	9·3059	9·3113	9·3167	9·3220	5	11	16	22	27	32	38	43	48
87	9·3274	9·3327	9·3381	9·3434	9·3488	9·3541	9·3595	9·3648	9·3702	9·3755	5	11	16	21	27	32	37	43	48
88	9·3808	9·3862	9·3915	9·3968	9·4021	9·4074	9·4128	9·4181	9·4234	9·4287	5	11	16	21	27	32	37	43	48
89	9·4340	9·4393	9·4446	9·4499	9·4552	9·4604	9·4657	9·4710	9·4763	9·4816	5	11	16	21	26	32	37	42	48
90	9·4868	9·4921	9·4974	9·5026	9·5079	9·5131	9·5184	9·5237	9·5289	9·5341	5	11	16	21	26	32	37	42	47
91	9·5394	9·5446	9·5499	9·5551	9·5603	9·5656	9·5708	9·5760	9·5812	9·5864	5	10	16	21	26	31	37	42	47
92	9·5917	9·5969	9·6021	9·6073	9·6125	9·6177	9·6229	9·6281	9·6333	9·6385	5	10	16	21	26	31	36	42	47
93	9·6437	9·6488	9·6540	9·6592	9·6644	9·6695	9·6747	9·6799	9·6850	9·6902	5	10	16	21	26	31	36	41	47
94	9·6954	9·7005	9·7057	9·7108	9·7160	9·7211	9·7263	9·7314	9·7365	9·7417	5	10	15	21	26	31	36	41	46
95	9·7468	9·7519	9·7570	9·7622	9·7673	9·7724	9·7775	9·7826	9·7877	9·7929	5	10	15	20	26	31	36	41	46
96	9·7980	9·8031	9·8082	9·8133	9·8184	9·8234	9·8285	9·8336	9·8387	9·8438	5	10	15	20	25	31	36	41	46
97	9·8489	9·8539	9·8590	9·8641	9·8691	9·9742	9·8793	9·8843	9·8894	9·8944	5	10	15	20	25	30	35	41	46
98	9·8995	9·9045	9·9096	9·9146	9·9197	9·9247	9·9298	9·9348	9·9398	9·9448	5	10	15	20	25	30	35	40	45
99	9·9499	9·9549	9·9599	9·9649	9·9700	9·9750	9·9800	9·9850	9·9900	9·9950	5	10	15	20	25	30	35	40	45

Table 11. Reciprocals $\frac{1}{x}$

										SUBTRACT								
x	0	1	2	3	4	5	6	7	8	9	1	2	3	4	5	6	7	8	9
1·0	1·0000	0·9901	0·9804	0·9709	0·9615	0·9524	0·9434	0·9346	0·9259	0·9174	9	18	28	37	46	55	64	73	82
1·1	0·9091	0·9009	0·8929	0·8850	0·8772	0·8696	0·8621	0·8547	0·8475	0·8403	8	15	23	31	38	46	54	61	69
1·2	0·8333	0·8264	0·8197	0·8130	0·8065	0·8000	0·7937	0·7874	0·7813	0·7752	6	13	19	26	32	39	45	52	58
1·3	0·7692	0·7634	0·7576	0·7519	0·7463	0·7407	0·7353	0·7299	0·7246	0·7194	6	11	17	22	28	33	39	44	50
1·4	0·7143	0·7092	0·7042	0·6993	0·6944	0·6897	0·6849	0·6803	0·6757	0·6711	5	10	14	19	24	29	34	38	43
1·5	0·6667	0·6623	0·6579	0·6536	0·6494	0·6452	0·6410	0·6369	0·6329	0·6289	4	8	13	17	21	25	29	34	38
1·6	0·6250	0·6211	0·6173	0·6135	0·6098	0·6061	0·6024	0·5988	0·5952	0·5917	4	7	11	15	19	22	26	30	33
1·7	0·5882	0·5848	0·5814	0·5780	0·5747	0·5714	0·5682	0·5650	0·5618	0·5587	3	7	10	13	16	20	23	26	30
1·8	0·5556	0·5525	0·5495	0·5464	0·5435	0·5405	0·5376	0·5348	0·5319	0·5291	3	6	9	12	15	18	21	24	26
1·9	0·5263	0·5236	0·5208	0·5181	0·5155	0·5128	0·5102	0·5076	0·5051	0·5025	3	5	8	11	13	16	19	21	24
2·0	0·5000	0·4975	0·4950	0·4926	0·4902	0·4878	0·4854	0·4831	0·4808	0·4785	2	5	7	10	12	14	17	19	22
2·1	0·4762	0·4739	0·4717	0·4695	0·4673	0·4651	0·4630	0·4608	0·4587	0·4566	2	4	7	9	11	13	15	17	20
2·2	0·4545	0·4525	0·4505	0·4484	0·4464	0·4444	0·4425	0·4405	0·4386	0·4367	2	4	6	8	10	12	14	16	18
2·3	0·4348	0·4329	0·4310	0·4292	0·4274	0·4255	0·4237	0·4219	0·4202	0·4184	2	4	5	7	9	11	13	15	16
2·4	0·4167	0·4149	0·4132	0·4115	0·4098	0·4082	0·4065	0·4049	0·4032	0·4016	2	3	5	7	8	10	12	13	15
2·5	0·4000	0·3984	0·3968	0·3953	0·3937	0·3922	0·3906	0·3891	0·3876	0·3861	2	3	5	6	8	9	11	12	14
2·6	0·3846	0·3831	0·3817	0·3802	0·3788	0·3774	0·3759	0·3745	0·3731	0·3717	1	3	4	6	7	9	10	11	13
2·7	0·3704	0·3690	0·3676	0·3663	0·3650	0·3636	0·3623	0·3610	0·3597	0·3584	1	3	4	5	7	8	9	11	12
2·8	0·3571	0·3559	0·3546	0·3534	0·3521	0·3509	0·3497	0·3484	0·3472	0·3460	1	2	4	5	6	7	9	10	11
2·9	0·3448	0·3436	0·3425	0·3413	0·3401	0·3390	0·3378	0·3367	0·3356	0·3344	1	2	3	5	6	7	8	9	10
3·0	0·3333	0·3322	0·3311	0·3300	0·3289	0·3279	0·3268	0·3257	0·3247	0·3236	1	2	3	4	5	6	8	9	10
3·1	0·3226	0·3215	0·3205	0·3195	0·3185	0·3175	0·3165	0·3155	0·3145	0·3135	1	2	3	4	5	6	7	8	9
3·2	0·3125	0·3115	0·3106	0·3096	0·3086	0·3077	0·3067	0·3058	0·3049	0·3040	1	2	3	4	5	6	7	8	9
3·3	0·3030	0·3021	0·3012	0·3003	0·2994	0·2985	0·2976	0·2967	0·2959	0·2950	1	2	3	4	4	5	6	7	8
3·4	0·2941	0·2933	0·2924	0·2915	0·2907	0·2899	0·2890	0·2882	0·2874	0·2865	1	2	3	3	4	5	6	7	8
3·5	0·2857	0·2849	0·2841	0·2833	0·2825	0·2817	0·2809	0·2801	0·2793	0·2786	1	2	2	3	4	5	6	6	7
3·6	0·2778	0·2770	0·2762	0·2755	0·2747	0·2740	0·2732	0·2725	0·2717	0·2710	1	2	2	3	4	5	5	6	7
3·7	0·2703	0·2695	0·2688	0·2681	0·2674	0·2667	0·2660	0·2653	0·2646	0·2639	1	1	2	3	4	4	5	6	6
3·8	0·2632	0·2625	0·2618	0·2611	0·2604	0·2597	0·2591	0·2584	0·2577	0·2571	1	1	2	3	3	4	5	5	6
3·9	0·2564	0·2558	0·2551	0·2545	0·2538	0·2532	0·2525	0·2519	0·2513	0·2506	1	1	2	3	3	4	5	5	6
4·0	0·2500	0·2494	0·2488	0·2481	0·2475	0·2469	0·2463	0·2457	0·2451	0·2445	1	1	2	2	3	4	4	5	6
4·1	0·2439	0·2433	0·2427	0·2421	0·2415	0·2410	0·2404	0·2398	0·2392	0·2387	1	1	2	2	3	3	4	5	5
4·2	0·2381	0·2375	0·2370	0·2364	0·2358	0·2353	0·2347	0·2342	0·2336	0·2331	1	1	2	2	3	3	4	4	5
4·3	0·2326	0·2320	0·2315	0·2309	0·2304	0·2299	0·2294	0·2288	0·2283	0·2278	1	1	2	2	3	3	4	4	5
4·4	0·2273	0·2268	0·2262	0·2257	0·2252	0·2247	0·2242	0·2237	0·2232	0·2227	1	1	2	2	3	3	4	4	5
4·5	0·2222	0·2217	0·2212	0·2208	0·2203	0·2198	0·2193	0·2188	0·2183	0·2179	0	1	1	2	2	3	3	4	4
4·6	0·2174	0·2169	0·2165	0·2160	0·2155	0·2151	0·2146	0·2141	0·2137	0·2132	0	1	1	2	2	3	3	4	4
4·7	0·2128	0·2123	0·2119	0·2114	0·2110	0·2105	0·2101	0·2096	0·2092	0·2088	0	1	1	2	2	3	3	4	4
4·8	0·2083	0·2079	0·2075	0·2070	0·2066	0·2062	0·2058	0·2053	0·2049	0·2045	0	1	1	2	2	3	3	3	4
4·9	0·2041	0·2037	0·2033	0·2028	0·2024	0·2020	0·2016	0·2012	0·2008	0·2004	0	1	1	2	2	2	3	3	4
5·0	0·2000	0·1996	0·1992	0·1988	0·1984	0·1980	0·1976	0·1972	0·1969	0·1965	0	1	1	2	2	2	3	3	4
5·1	0·1961	0·1957	0·1953	0·1949	0·1946	0·1942	0·1938	0·1934	0·1931	0·1927	0	1	1	2	2	2	3	3	3
5·2	0·1923	0·1919	0·1916	0·1912	0·1908	0·1905	0·1901	0·1898	0·1894	0·1890	0	1	1	1	2	2	3	3	3
5·3	0·1887	0·1883	0·1880	0·1876	0·1873	0·1869	0·1866	0·1862	0·1859	0·1855	0	1	1	1	2	2	2	3	3
5·4	0·1852	0·1848	0·1845	0·1842	0·1838	0·1835	0·1832	0·1828	0·1825	0·1821	0	1	1	1	2	2	2	3	3

*

Reciprocals

$$\frac{1}{x}$$

											SUBTRACT								
x	0	1	2	3	4	5	6	7	8	9	1	2	3	4	5	6	7	8	9
5·5	0·1818	0·1815	0·1812	0·1808	0·1805	0·1802	0·1799	0·1795	0·1792	0·1789	0	1	1	1	2	2	2	3	3
5·6	0·1786	0·1783	0·1779	0·1776	0·1773	0·1770	0·1767	0·1764	0·1761	0·1757	0	1	1	1	2	2	2	3	3
5·7	0·1754	0·1751	0·1748	0·1745	0·1742	0·1739	0·1736	0·1733	0·1730	0·1727	0	1	1	1	2	2	2	2	3
5·8	0·1724	0·1721	0·1718	0·1715	0·1712	0·1709	0·1706	0·1704	0·1701	0·1698	0	1	1	1	1	2	2	2	3
5·9	0·1695	0·1692	0·1689	0·1686	0·1684	0·1681	0·1678	0·1675	0·1672	0·1669	0	1	1	1	1	2	2	2	3
6·0	0·1667	0·1664	0·1661	0·1658	0·1656	0·1653	0·1650	0·1647	0·1645	0·1642	0	1	1	1	1	2	2	2	2
6·1	0·1639	0·1637	0·1634	0·1631	0·1629	0·1626	0·1623	0·1621	0·1618	0·1616	0	1	1	1	1	2	2	2	2
6·2	0·1613	0·1610	0·1608	0·1605	0·1603	0·1600	0·1597	0·1595	0·1592	0·1590	0	1	1	1	1	2	2	2	2
6·3	0·1587	0·1585	0·1582	0·1580	0·1577	0·1575	0·1572	0·1570	0·1567	0·1565	0	0	1	1	1	1	2	2	2
6·4	0·1563	0·1560	0·1558	0·1555	0·1553	0·1550	0·1548	0·1546	0·1543	0·1541	0	0	1	1	1	1	2	2	2
6·5	0·1538	0·1536	0·1534	0·1531	0·1529	0·1527	0·1524	0·1522	0·1520	0·1517	0	0	1	1	1	1	2	2	2
6·6	0·1515	0·1513	0·1511	0·1508	0·1506	0·1504	0·1502	0·1499	0·1497	0·1495	0	0	1	1	1	1	2	2	2
6·7	0·1493	0·1490	0·1488	0·1486	0·1484	0·1481	0·1479	0·1477	0·1475	0·1473	0	0	1	1	1	1	2	2	2
6·8	0·1471	0·1468	0·1466	0·1464	0·1462	0·1460	0·1458	0·1456	0·1453	0·1451	0	0	1	1	1	1	1	2	2
6·9	0·1449	0·1447	0·1445	0·1443	0·1441	0·1439	0·1437	0·1435	0·1433	0·1431	0	0	1	1	1	1	1	2	2
7·0	0·1429	0·1427	0·1425	0·1422	0·1420	0·1418	0·1416	0·1414	0·1412	0·1410	0	0	1	1	1	1	1	2	2
7·1	0·1408	0·1406	0·1404	0·1403	0·1401	0·1399	0·1397	0·1395	0·1393	0·1391	0	0	1	1	1	1	1	2	2
7·2	0·1389	0·1387	0·1385	0·1383	0·1381	0·1379	0·1377	0·1376	0·1374	0·1372	0	0	1	1	1	1	1	2	2
7·3	0·1370	0·1368	0·1366	0·1364	0·1362	0·1361	0·1359	0·1357	0·1355	0·1353	0	0	1	1	1	1	1	1	2
7·4	0·1351	0·1350	0·1348	0·1346	0·1344	0·1342	0·1340	0·1339	0·1337	0·1335	0	0	1	1	1	1	1	1	2
7·5	0·1333	0·1332	0·1330	0·1328	0·1326	0·1325	0·1323	0·1321	0·1319	0·1318	0	0	1	1	1	1	1	1	2
7·6	0·1316	0·1314	0·1312	0·1311	0·1309	0·1307	0·1305	0·1304	0·1302	0·1300	0	0	1	1	1	1	1	1	2
7·7	0·1299	0·1297	0·1295	0·1294	0·1292	0·1290	0·1289	0·1287	0·1285	0·1284	0	0	1	1	1	1	1	1	2
7·8	0·1282	0·1280	0·1279	0·1277	0·1276	0·1274	0·1272	0·1271	0·1269	0·1267	0	0	0	1	1	1	1	1	1
7·9	0·1266	0·1264	0·1263	0·1261	0·1259	0·1258	0·1256	0·1255	0·1253	0·1252	0	0	0	1	1	1	1	1	1
8·0	0·1250	0·1248	0·1247	0·1245	0·1244	0·1242	0·1241	0·1239	0·1238	0·1236	0	0	0	1	1	1	1	1	1
8·1	0·1235	0·1233	0·1232	0·1230	0·1229	0·1227	0·1225	0·1224	0·1222	0·1221	0	0	0	1	1	1	1	1	1
8·2	0·1220	0·1218	0·1217	0·1215	0·1214	0·1212	0·1211	0·1209	0·1208	0·1206	0	0	0	1	1	1	1	1	1
8·3	0·1205	0·1203	0·1202	0·1200	0·1199	0·1198	0·1196	0·1195	0·1193	0·1192	0	0	0	1	1	1	1	1	1
8·4	0·1190	0·1189	0·1188	0·1186	0·1185	0·1183	0·1182	0·1181	0·1179	0·1178	0	0	0	1	1	1	1	1	1
8·5	0·1176	0·1175	0·1174	0·1172	0·1171	0·1170	0·1168	0·1167	0·1166	0·1164	0	0	0	1	1	1	1	1	1
8·6	0·1163	0·1161	0·1160	0·1159	0·1157	0·1156	0·1155	0·1153	0·1152	0·1151	0	0	0	1	1	1	1	1	1
8·7	0·1149	0·1148	0·1147	0·1145	0·1144	0·1143	0·1142	0·1140	0·1139	0·1138	0	0	0	1	1	1	1	1	1
8·8	0·1136	0·1135	0·1134	0·1133	0·1131	0·1130	0·1129	0·1127	0·1126	0·1125	0	0	0	1	1	1	1	1	1
8·9	0·1124	0·1122	0·1121	0·1120	0·1119	0·1117	0·1116	0·1115	0·1114	0·1112	0	0	0	1	1	1	1	1	1
9·0	0·1111	0·1110	0·1109	0·1107	0·1106	0·1105	0·1104	0·1103	0·1101	0·1100	0	0	0	0	1	1	1	1	1
9·1	0·1099	0·1098	0·1096	0·1095	0·1094	0·1093	0·1092	0·1091	0·1089	0·1088	0	0	0	0	1	1	1	1	1
9·2	0·1087	0·1086	0·1085	0·1083	0·1082	0·1081	0·1080	0·1079	0·1078	0·1076	0	0	0	0	1	1	1	1	1
9·3	0·1075	0·1074	0·1073	0·1072	0·1071	0·1070	0·1068	0·1067	0·1066	0·1065	0	0	0	0	1	1	1	1	1
9·4	0·1064	0·1063	0·1062	0·1060	0·1059	0·1058	0·1057	0·1056	0·1055	0·1054	0	0	0	0	1	1	1	1	1
9·5	0·1053	0·1052	0·1050	0·1049	0·1048	0·1047	0·1046	0·1045	0·1044	0·1043	0	0	0	0	1	1	1	1	1
9·6	0·1042	0·1041	0·1040	0·1038	0·1037	0·1036	0·1035	0·1034	0·1033	0·1032	0	0	0	0	1	1	1	1	1
9·7	0·1031	0·1030	0·1029	0·1028	0·1027	0·1026	0·1025	0·1024	0·1022	0·1021	0	0	0	0	1	1	1	1	1
9·8	0·1020	0·1019	0·1018	0·1017	0·1016	0·1015	0·1014	0·1013	0·1012	0·1011	0	0	0	0	1	1	1	1	1
9·9	0·1010	0·1009	0·1008	0·1007	0·1006	0·1005	0·1004	0·1003	0·1002	0·1001	0	0	0	0	1	1	1	1	1

Table 12. Natural Logarithms ln x

x	0	1	2	3	4	5	6	7	8	9	1	2	3	4	5	6	7	8	9
1·0	0·0000	0·0100	0·0198	0·0296	0·0392	0·0488	0·0583	0·0677	0·0770	0·0862	10	19	29	38	48	58	67	77	86
1·1	0·0953	0·1044	0·1133	0·1222	0·1310	0·1398	0·1484	0·1570	0·1655	0·1740	9	18	26	35	44	52	61	70	79
1·2	0·1823	0·1906	0·1989	0·2070	0·2151	0·2231	0·2311	0·2390	0·2469	0·2546	8	16	24	32	40	48	56	64	72
1·3	0·2624	0·2700	0·2776	0·2852	0·2927	0·3001	0·3075	0·3148	0·3221	0·3293	7	15	22	30	37	45	52	60	67
1·4	0·3365	0·3436	0·3507	0·3577	0·3646	0·3716	0·3784	0·3853	0·3920	0·3988	7	14	21	28	35	42	48	55	62
1·5	0·4055	0·4121	0·4187	0·4253	0·4318	0·4383	0·4447	0·4511	0·4574	0·4637	6	13	19	26	32	39	45	52	58
1·6	0·4700	0·4762	0·4824	0·4886	0·4947	0·5008	0·5068	0·5128	0·5188	0·5247	6	12	18	24	30	37	43	49	55
1·7	0·5306	0·5365	0·5423	0·5481	0·5539	0·5596	0·5653	0·5710	0·5766	0·5822	6	11	17	23	29	34	40	46	52
1·8	0·5878	0·5933	0·5988	0·6043	0·6098	0·6152	0·6206	0·6259	0·6313	0·6366	5	11	16	22	27	33	38	43	49
1·9	0·6419	0·6471	0·6523	0·6575	0·6627	0·6678	0·6729	0·6780	0·6831	0·6881	5	10	15	21	26	31	36	41	46
2·0	0·6931	0·6981	0·7031	0·7080	0·7130	0·7178	0·7227	0·7276	0·7324	0·7372	5	10	15	20	24	29	34	39	44
2·1	0·7419	0·7467	0·7514	0·7561	0·7608	0·7655	0·7701	0·7747	0·7793	0·7839	5	9	14	19	23	28	33	37	42
2·2	0·7885	0·7930	0·7975	0·8020	0·8065	0·8109	0·8154	0·8198	0·8242	0·8286	4	9	13	18	22	27	31	36	40
2·3	0·8329	0·8372	0·8416	0·8459	0·8502	0·8544	0·8587	0·8629	0·8671	0·8713	4	9	13	17	21	26	30	34	38
2·4	0·8755	0·8796	0·8838	0·8879	0·8920	0·8961	0·9002	0·9042	0·9083	0·9123	4	8	12	16	20	25	29	33	37
2·5	0·9163	0·9203	0·9243	0·9282	0·9322	0·9361	0·9400	0·9439	0·9478	0·9517	4	8	12	16	20	24	28	31	35
2·6	0·9555	0·9594	0·9632	0·9670	0·9708	0·9746	0·9783	0·9821	0·9858	0·9895	4	8	11	15	19	23	26	30	34
2·7	0·9933	0·9969	1·0006	1·0043	1·0080	1·0116	1·0152	1·0188	1·0225	1·0260	4	7	11	15	18	22	26	29	33
2·8	1·0296	1·0332	1·0367	1·0403	1·0438	1·0473	1·0508	1·0543	1·0578	1·0613	4	7	11	14	18	21	25	28	32
2·9	1·0647	1·0682	1·0716	1·0750	1·0784	1·0818	1·0852	1·0886	1·0919	1·0953	3	7	10	14	17	20	24	27	31
3·0	1·0986	1·1019	1·1053	1·1086	1·1119	1·1151	1·1184	1·1217	1·1249	1·1282	3	7	10	13	16	20	23	26	30
3·1	1·1314	1·1346	1·1378	1·1410	1·1442	1·1474	1·1506	1·1537	1·1569	1·1600	3	6	10	13	16	19	22	25	29
3·2	1·1632	1·1663	1·1694	1·1725	1·1756	1·1787	1·1817	1·1848	1·1878	1·1909	3	6	9	12	15	19	22	25	28
3·3	1·1939	1·1969	1·2000	1·2030	1·2060	1·2090	1·2119	1·2149	1·2179	1·2208	3	6	9	12	15	18	21	24	27
3·4	1·2238	1·2267	1·2296	1·2326	1·2355	1·2384	1·2413	1·2442	1·2470	1·2499	3	6	9	12	15	17	20	23	26
3·5	1·2528	1·2556	1·2585	1·2613	1·2641	1·2669	1·2698	1·2726	1·2754	1·2782	3	6	8	11	14	17	20	23	25
3·6	1·2809	1·2837	1·2865	1·2892	1·2920	1·2947	1·2975	1·3002	1·3029	1·3056	3	6	8	11	14	16	19	22	25
3·7	1·3083	1·3110	1·3137	1·3164	1·3191	1·3218	1·3244	1·3271	1·3297	1·3324	3	5	8	11	13	16	19	21	24
3·8	1·3350	1·3376	1·3403	1·3429	1·3455	1·3481	1·3507	1·3533	1·3558	1·3584	3	5	8	10	13	16	18	21	23
3·9	1·3610	1·3635	1·3661	1·3686	1·3712	1·3737	1·3762	1·3788	1·3813	1·3838	3	5	8	10	13	15	18	20	23
4·0	1·3863	1·3888	1·3913	1·3938	1·3962	1·3987	1·4012	1·4036	1·4061	1·4085	2	5	7	10	12	15	17	20	22
4·1	1·4110	1·4134	1·4159	1·4183	1·4207	1·4231	1·4255	1·4279	1·4303	1·4327	2	5	7	10	12	14	17	19	22
4·2	1·4351	1·4375	1·4398	1·4422	1·4446	1·4469	1·4493	1·4516	1·4540	1·4563	2	5	7	9	12	14	16	19	21
4·3	1·4586	1·4609	1·4633	1·4656	1·4679	1·4702	1·4725	1·4748	1·4770	1·4793	2	5	7	9	12	14	16	18	21
4·4	1·4816	1·4839	1·4861	1·4884	1·4907	1·4929	1·4951	1·4974	1·4996	1·5019	2	5	7	9	11	14	16	18	20
4·5	1·5041	1·5063	1·5085	1·5107	1·5129	1·5151	1·5173	1·5195	1·5217	1·5239	2	4	7	9	11	13	15	18	20
4·6	1·5261	1·5282	1·5304	1·5326	1·5347	1·5369	1·5390	1·5412	1·5433	1·5454	2	4	6	9	11	13	15	17	19
4·7	1·5476	1·5497	1·5518	1·5539	1·5560	1·5581	1·5602	1·5623	1·5644	1·5665	2	4	6	8	11	13	15	17	19
4·8	1·5686	1·5707	1·5728	1·5748	1·5769	1·5790	1·5810	1·5831	1·5851	1·5872	2	4	6	8	10	12	14	17	19
4·9	1·5892	1·5913	1·5933	1·5953	1·5974	1·5994	1·6014	1·6034	1·6054	1·6074	2	4	6	8	10	12	14	16	18
5·0	1·6094	1·6114	1·6134	1·6154	1·6174	1·6194	1·6214	1·6233	1·6253	1·6273	2	4	6	8	10	12	14	16	18
5·1	1·6292	1·6312	1·6332	1·6351	1·6371	1·6390	1·6409	1·6429	1·6448	1·6467	2	4	6	8	10	12	14	16	18
5·2	1·6487	1·6506	1·6525	1·6544	1·6563	1·6582	1·6601	1·6620	1·6639	1·6658	2	4	6	8	10	11	13	15	17
5·3	1·6677	1·6696	1·6715	1·6734	1·6752	1·6771	1·6790	1·6808	1·6827	1·6845	2	4	6	7	9	11	13	15	17
5·4	1·6864	1·6882	1·6901	1·6919	1·6938	1·6956	1·6974	1·6993	1·7011	1·7029	2	4	6	7	9	11	13	15	17

ln 10^x

x	1	2	3	4	5	6	7	8	9	10
ln10^x	2·3026	4·6052	6·9078	9·2103	11·5129	13·8155	16·1181	18·4207	20·7233	23·0259

Natural Logarithms $\ln x$

x	0	1	2	3	4	5	6	7	8	9	1	2	3	4	5	6	7	8	9
5·5	1·7047	1·7066	1·7084	1·7102	1·7120	1·7138	1·7156	1·7174	1·7192	1·7210	2	4	5	7	9	11	13	14	16
5·6	1·7228	1·7246	1·7263	1·7281	1·7299	1·7317	1·7334	1·7352	1·7370	1·7387	2	4	5	7	9	11	12	14	16
5·7	1·7405	1·7422	1·7440	1·7457	1·7475	1·7492	1·7509	1·7527	1·7544	1·7561	2	3	5	7	9	10	12	14	16
5·8	1·7579	1·7596	1·7613	1·7630	1·7647	1·7664	1·7682	1·7699	1·7716	1·7733	2	3	5	7	9	10	12	14	15
5·9	1·7750	1·7766	1·7783	1·7800	1·7817	1·7834	1·7851	1·7867	1·7884	1·7901	2	3	5	7	8	10	12	13	15
6·0	1·7918	1·7934	1·7951	1·7967	1·7984	1·8001	1·8017	1·8034	1·8050	1·8066	2	3	5	7	8	10	12	13	15
6·1	1·8083	1·8099	1·8116	1·8132	1·8148	1·8165	1·8181	1·8197	1·8213	1·8229	2	3	5	7	8	10	11	13	15
6·2	1·8246	1·8262	1·8278	1·8294	1·8310	1·8326	1·8342	1·8358	1·8374	1·8390	2	3	5	6	8	10	11	13	14
6·3	1·8406	1·8421	1·8437	1·8453	1·8469	1·8485	1·8500	1·8516	1·8532	1·8547	2	3	5	6	8	9	11	13	14
6·4	1·8563	1·8579	1·8594	1·8610	1·8625	1·8641	1·8656	1·8672	1·8687	1·8703	2	3	5	6	8	9	11	12	14
6·5	1·8718	1·8733	1·8749	1·8764	1·8779	1·8795	1·8810	1·8825	1·8840	1·8856	2	3	5	6	8	9	11	12	14
6·6	1·8871	1·8886	1·8901	1·8916	1·8931	1·8946	1·8961	1·8976	1·8991	1·9006	2	3	5	6	8	9	11	12	14
6·7	1·9021	1·9036	1·9051	1·9066	1·9081	1·9095	1·9110	1·9125	1·9140	1·9155	1	3	4	6	7	9	10	12	13
6·8	1·9169	1·9184	1·9199	1·9213	1·9228	1·9242	1·9257	1·9272	1·9286	1·9301	1	3	4	6	7	9	10	12	13
6·9	1·9315	1·9330	1·9344	1·9359	1·9373	1·9387	1·9402	1·9416	1·9430	1·9445	1	3	4	6	7	9	10	12	13
7·0	1·9459	1·9473	1·9488	1·9502	1·9516	1·9530	1·9544	1·9559	1·9573	1·9587	1	3	4	6	7	9	10	11	13
7·1	1·9601	1·9615	1·9629	1·9643	1·9657	1·9671	1·9685	1·9699	1·9713	1·9727	1	3	4	6	7	8	10	11	13
7·2	1·9741	1·9755	1·9769	1·9782	1·9796	1·9810	1·9824	1·9838	1·9851	1·9865	1	3	4	6	7	8	10	11	12
7·3	1·9879	1·9892	1·9906	1·9920	1·9933	1·9947	1·9961	1·9974	1·9988	2·0001	1	3	4	5	7	8	10	11	12
7·4	2·0015	2·0028	2·0042	2·0055	2·0069	2·0082	2·0096	2·0109	2·0122	2·0136	1	3	4	5	7	8	9	11	12
7·5	2·0149	2·0162	2·0176	2·0189	2·0202	2·0215	2·0229	2·0242	2·0255	2·0268	1	3	4	5	7	8	9	11	12
7·6	2·0281	2·0295	2·0308	2·0321	2·0334	2·0347	2·0360	2·0373	2·0386	2·0399	1	3	4	5	7	8	9	10	12
7·7	2·0412	2·0425	2·0438	2·0451	2·0464	2·0477	2·0490	2·0503	2·0516	2·0528	1	3	4	5	6	8	9	10	12
7·8	2·0541	2·0554	2·0567	2·0580	2·0592	2·0605	2·0618	2·0631	2·0643	2·0656	1	3	4	5	6	8	9	10	11
7·9	2·0669	2·0681	2·0694	2·0707	2·0719	2·0732	2·0744	2·0757	2·0769	2·0782	1	3	4	5	6	8	9	10	11
8·0	2·0794	2·0807	2·0819	2·0832	2·0844	2·0857	2·0869	2·0882	2·0894	2·0906	1	2	4	5	6	7	9	10	11
8·1	2·0919	2·0931	2·0943	2·0956	2·0968	2·0980	2·0992	2·1005	2·1017	2·1029	1	2	4	5	6	7	9	10	11
8·2	2·1041	2·1054	2·1066	2·1078	2·1090	2·1102	2·1114	2·1126	2·1138	2·1151	1	2	4	5	6	7	8	10	11
8·3	2·1163	2·1175	2·1187	2·1199	2·1211	2·1223	2·1235	2·1247	2·1258	2·1270	1	2	4	5	6	7	8	10	11
8·4	2·1282	2·1294	2·1306	2·1318	2·1330	2·1342	2·1354	2·1365	2·1377	2·1389	1	2	4	5	6	7	8	9	11
8·5	2·1401	2·1412	2·1424	2·1436	2·1448	2·1459	2·1471	2·1483	2·1494	2·1506	1	2	4	5	6	7	8	9	11
8·6	2·1518	2·1529	2·1541	2·1552	2·1564	2·1576	2·1587	2·1599	2·1610	2·1622	1	2	3	5	6	7	8	9	10
8·7	2·1633	2·1645	2·1656	2·1668	2·1679	2·1691	2·1702	2·1713	2·1725	2·1736	1	2	3	5	6	7	8	9	10
8·8	2·1748	2·1759	2·1770	2·1782	2·1793	2·1804	2·1815	2·1827	2·1838	2·1849	1	2	3	5	6	7	8	9	10
8·9	2·1861	2·1872	2·1883	2·1894	2·1905	2·1917	2·1928	2·1939	2·1950	2·1961	1	2	3	4	6	7	8	9	10
9·0	2·1972	2·1983	2·1994	2·2006	2·2017	2·2028	2·2039	2·2050	2·2061	2·2072	1	2	3	4	6	7	8	9	10
9·1	2·2083	2·2094	2·2105	2·2116	2·2127	2·2138	2·2148	2·2159	2·2170	2·2181	1	2	3	4	5	7	8	9	10
9·2	2·2192	2·2203	2·2214	2·2225	2·2235	2·2246	2·2257	2·2268	2·2279	2·2289	1	2	3	4	5	6	8	9	10
9·3	2·2300	2·2311	2·2322	2·2332	2·2343	2·2354	2·2364	2·2375	2·2386	2·2396	1	2	3	4	5	6	7	9	10
9·4	2·2407	2·2418	2·2428	2·2439	2·2450	2·2460	2·2471	2·2481	2·2492	2·2502	1	2	3	4	5	6	7	8	10
9·5	2·2513	2·2523	2·2534	2·2544	2·2555	2·2565	2·2576	2·2586	2·2597	2·2607	1	2	3	4	5	6	7	8	9
9·6	2·2618	2·2628	2·2638	2·2649	2·2659	2·2670	2·2680	2·2690	2·2701	2·2711	1	2	3	4	5	6	7	8	9
9·7	2·2721	2·2732	2·2742	2·2752	2·2762	2·2773	2·2783	2·2793	2·2803	2·2814	1	2	3	4	5	6	7	8	9
9·8	2·2824	2·2834	2·2844	2·2854	2·2865	2·2875	2·2885	2·2895	2·2905	2·2915	1	2	3	4	5	6	7	8	9
9·9	2·2925	2·2935	2·2946	2·2956	2·2966	2·2976	2·2986	2·2996	2·3006	2·3016	1	2	3	4	5	6	7	8	9

For further values use the table at the foot of p.30;
e.g., $4750 = 10^3 \times 4{\cdot}750$; $\ln 4750 = 6{\cdot}9078 + 1{\cdot}5581 = 8{\cdot}4659$.

Table 13. e^x

x	·00	·01	·02	·03	·04	·05	·06	·07	·08	·09
0·0	1·0000	1·0100	1·0202	1·0305	1·0408	1·0513	1·0618	1·0725	1·0833	1·0942
0·1	1·1052	1·1163	1·1275	1·1388	1·1503	1·1618	1·1735	1·1853	1·1972	1·2092
0·2	1·2214	1·2337	1·2461	1·2586	1·2712	1·2840	1·2969	1·3100	1·3231	1·3364
0·3	1·3499	1·3634	1·3771	1·3910	1·4049	1·4191	1·4333	1·4477	1·4623	1·4770
0·4	1·4918	1·5068	1·5220	1·5373	1·5527	1·5683	1·5841	1·6000	1·6161	1·6323
0·5	1·6487	1·6653	1·6820	1·6989	1·7160	1·7333	1·7507	1·7683	1·7860	1·8040
0·6	1·8221	1·8404	1·8589	1·8776	1·8965	1·9155	1·9348	1·9542	1·9739	1·9937
0·7	2·0138	2·0340	2·0544	2·0751	2·0959	2·1170	2·1383	2·1598	2·1815	2·2034
0·8	2·2255	2·2479	2·2705	2·2933	2·3164	2·3396	2·3632	2·3869	2·4109	2·4351
0·9	2·4596	2·4843	2·5093	2·5345	2·5600	2·5857	2·6117	2·6379	2·6645	2·6912
1·0	2·7183	2·7456	2·7732	2·8011	2·8292	2·8576	2·8864	2·9154	2·9447	2·9743
1·1	3·0042	3·0344	3·0649	3·0957	3·1268	3·1582	3·1899	3·2220	3·2544	3·2871
1·2	3·3201	3·3535	3·3872	3·4212	3·4556	3·4903	3·5254	3·5609	3·5966	3·6328
1·3	3·6693	3·7062	3·7434	3·7810	3·8190	3·8574	3·8962	3·9354	3·9749	4·0148
1·4	4·0552	4·0960	4·1371	4·1787	4·2207	4·2631	4·3060	4·3492	4·3929	4·4371
1·5	4·4817	4·5267	4·5722	4·6182	4·6646	4·7115	4·7588	4·8066	4·8550	4·9037
1·6	4·9530	5·0028	5·0531	5·1039	5·1552	5·2070	5·2593	5·3122	5·3656	5·4195
1·7	5·4739	5·5290	5·5845	5·6407	5·6973	5·7546	5·8124	5·8709	5·9299	5·9895
1·8	6·0496	6·1104	6·1719	6·2339	6·2965	6·3598	6·4237	6·4883	6·5535	6·6194
1·9	6·6859	6·7531	6·8210	6·8895	6·9587	7·0287	7·0993	7·1707	7·2427	7·3155
2·0	7·3891	7·4633	7·5383	7·6141	7·6906	7·7679	7·8460	7·9248	8·0045	8·0849
2·1	8·1662	8·2482	8·3311	8·4149	8·4994	8·5849	8·6711	8·7583	8·8463	8·9352
2·2	9·0250	9·1157	9·2073	9·2999	9·3933	9·4877	9·5831	9·6794	9·7767	9·8749
2·3	9·9742	10·074	10·176	10·278	10·381	10·486	10·591	10·697	10·805	10·914
2·4	11·023	11·134	11·246	11·359	11·473	11·588	11·705	11·822	11·941	12·061
2·5	12·183	12·305	12·429	12·554	12·680	12·807	12·936	13·066	13·197	13·330
2·6	13·464	13·599	13·736	13·874	14·013	14·154	14·296	14·440	14·585	14·732
2·7	14·880	15·029	15·180	15·333	15·487	15·643	15·800	15·959	16·119	16·281
2·8	16·445	16·610	16·777	16·946	17·116	17·288	17·462	17·637	17·814	17·993
2·9	18·174	18·357	18·541	18·728	18·916	19·106	19·298	19·492	19·688	19·886
3·0	20·086	20·287	20·491	20·697	20·905	21·115	21·328	21·542	21·758	21·977
3·1	22·198	22·421	22·646	22·874	23·104	23·336	23·571	23·808	24·047	24·288
3·2	24·533	24·779	25·028	25·280	25·534	25·790	26·050	26·311	26·576	26·843
3·3	27·113	27·385	27·660	27·938	28·219	28·503	28·789	29·079	29·371	29·666
3·4	29·964	30·265	30·569	30·877	31·187	31·500	31·817	32·137	32·460	32·786
3·5	33·115	33·448	33·784	34·124	34·467	34·813	35·163	35·517	35·874	36·234
3·6	36·598	36·966	37·338	37·713	38·092	38·475	38·861	39·252	39·646	40·045
3·7	40·447	40·854	41·264	41·679	42·098	42·521	42·948	43·380	43·816	44·256
3·8	44·701	45·150	45·604	46·063	46·525	46·993	47·465	47·942	48·424	48·911
3·9	49·402	49·899	50·400	50·907	51·419	51·935	52·457	52·985	53·517	54·055
4·0	54·598	55·147	55·701	56·261	56·826	57·397	57·974	58·557	59·146	59·740

e^x

x	·0	·1	·2	·3	·4	·5	·6	·7	·8	·9
4	54·598	60·340	66·686	73·700	81·451	90·017	99·484	109·95	121·51	134·29
5	148·41	164·02	181·27	200·34	221·41	244·69	270·43	298·87	330·30	365·04
6	403·43	445·86	492·75	544·57	601·84	665·14	735·09	812·41	897·85	992·27
7	1096·6	1212·0	1339·4	1480·3	1636·0	1808·0	1998·2	2208·3	2440·6	2697·3
8	2981·0	3294·5	3640·9	4023·9	4447·1	4914·8	5431·7	6002·9	6634·2	7332·0
9	8103·1	8955·3	9897·1	10938	12088	13360	14765	16318	18034	19930
10	22026	24343	26903	29733	32860	36316	40135	44356	49021	54176

For further values use the table at the foot of p.30;
e.g., $e^{14 \cdot 5} = e^{13 \cdot 8155} \times e^{\cdot 6845} = 10^6 \times 1 \cdot 983$

Table 14. e^{-x}

x	·00	·01	·02	·03	·04	·05	·06	·07	·08	·09
0·0	1·0000	0·9900	0·9802	0·9704	0·9608	0·9512	0·9418	0·9324	0·9231	0·9139
0·1	0·9048	0·8958	0·8869	0·8781	0·8694	0·8607	0·8521	0·8437	0·8353	0·8270
0·2	0·8187	0·8106	0·8025	0·7945	0·7866	0·7788	0·7711	0·7634	0·7558	0·7483
0·3	0·7408	0·7334	0·7261	0·7189	0·7118	0·7047	0·6977	0·6907	0·6839	0·6771
0·4	0·6703	0·6637	0·6570	0·6505	0·6440	0·6376	0·6313	0·6250	0·6188	0·6126
0·5	0·6065	0·6005	0·5945	0·5886	0·5827	0·5769	0·5712	0·5655	0·5599	0·5543
0·6	0·5488	0·5434	0·5379	0·5326	0·5273	0·5220	0·5169	0·5117	0·5066	0·5016
0·7	0·4966	0·4916	0·4868	0·4819	0·4771	0·4724	0·4677	0·4630	0·4584	0·4538
0·8	0·4493	0·4449	0·4404	0·4360	0·4317	0·4274	0·4232	0·4190	0·4148	0·4107
0·9	0·4066	0·4025	0·3985	0·3946	0·3906	0·3867	0·3829	0·3791	0·3753	0·3716
1·0	0·3679	0·3642	0·3606	0·3570	0·3535	0·3499	0·3465	0·3430	0·3396	0·3362
1·1	0·3329	0·3296	0·3263	0·3230	0·3198	0·3166	0·3135	0·3104	0·3073	0·3042
1·2	0·3012	0·2982	0·2952	0·2923	0·2894	0·2865	0·2837	0·2808	0·2780	0·2753
1·3	0·2725	0·2698	0·2671	0·2645	0·2618	0·2592	0·2567	0·2541	0·2516	0·2491
1·4	0·2466	0·2441	0·2417	0·2393	0·2369	0·2346	0·2322	0·2299	0·2276	0·2254
1·5	0·2231	0·2209	0·2187	0·2165	0·2144	0·2122	0·2101	0·2080	0·2060	0·2039
1·6	0·2019	0·1999	0·1979	0·1959	0·1940	0·1920	0·1901	0·1882	0·1864	0·1845
1·7	0·1827	0·1809	0·1791	0·1773	0·1755	0·1738	0·1720	0·1703	0·1686	0·1670
1·8	0·1653	0·1637	0·1620	0·1604	0·1588	0·1572	0·1557	0·1541	0·1526	0·1511
1·9	0·1496	0·1481	0·1466	0·1451	0·1437	0·1423	0·1409	0·1395	0·1381	0·1367
2·0	0·1353	0·1340	0·1327	0·1313	0·1300	0·1287	0·1275	0·1262	0·1249	0·1237
2·1	0·1225	0·1212	0·1200	0·1188	0·1177	0·1165	0·1153	0·1142	0·1130	0·1119
2·2	0·1108	0·1097	0·1086	0·1075	0·1065	0·1054	0·1044	0·1033	0·1023	0·1013
2·3	0·1003	0·0993	0·0983	0·0973	0·0963	0·0954	0·0944	0·0935	0·0926	0·0916
2·4	0·0907	0·0898	0·0889	0·0880	0·0872	0·0863	0·0854	0·0846	0·0837	0·0829
2·5	0·0821	0·0813	0·0805	0·0797	0·0789	0·0781	0·0773	0·0765	0·0758	0·0750
2·6	0·0743	0·0735	0·0728	0·0721	0·0714	0·0707	0·0699	0·0693	0·0686	0·0679
2·7	0·0672	0·0665	0·0659	0·0652	0·0646	0·0639	0·0633	0·0627	0·0620	0·0614
2·8	0·0608	0·0602	0·0596	0·0590	0·0584	0·0578	0·0573	0·0567	0·0561	0·0556
2·9	0·0550	0·0545	0·0539	0·0534	0·0529	0·0523	0·0518	0·0513	0·0508	0·0503
3·0	0·0498	0·0493	0·0488	0·0483	0·0478	0·0474	0·0469	0·0464	0·0460	0·0455
3·1	0·0450	0·0446	0·0442	0·0437	0·0433	0·0429	0·0424	0·0420	0·0416	0·0412
3·2	0·0408	0·0404	0·0400	0·0396	0·0392	0·0388	0·0384	0·0380	0·0376	0·0373
3·3	0·0369	0·0365	0·0362	0·0358	0·0354	0·0351	0·0347	0·0344	0·0340	0·0337
3·4	0·0334	0·0330	0·0327	0·0324	0·0321	0·0317	0·0314	0·0311	0·0308	0·0305
3·5	0·0302	0·0299	0·0296	0·0293	0·0290	0·0287	0·0284	0·0282	0·0279	0·0276
3·6	0·0273	0·0271	0·0268	0·0265	0·0263	0·0260	0·0257	0·0255	0·0252	0·0250
3·7	0·0247	0·0245	0·0242	0·0240	0·0238	0·0235	0·0233	0·0231	0·0228	0·0226
3·8	0·0224	0·0221	0·0219	0·0217	0·0215	0·0213	0·0211	0·0209	0·0207	0·0204
3·9	0·0202	0·0200	0·0198	0·0196	0·0194	0·0193	0·0191	0·0189	0·0187	0·0185
4·0	0·0183	0·0181	0·0180	0·0178	0·0176	0·0174	0·0172	0·0171	0·0169	0·0167

e^{-x}

x	·0	·1	·2	·3	·4	·5	·6	·7	·8	·9
4	·01832	·01657	·01500	·01357	·01228	·01111	·01005	·00910	·00823	·00745
5	·00674	·00610	·00552	·00499	·00452	·00409	·00370	·00335	·00303	·00274
6	·00248	·00224	·00203	·00184	·00166	·00150	·00136	·00123	·00111	·00101
7	·00091	·00083	·00075	·00068	·00061	·00055	·00050	·00045	·00041	·00037
8	·00034	·00030	·00027	·00025	·00022	·00020	·00018	·00017	·00015	·00014
9	·00012	·00011	·00010	·00009	·00008	·00007	·00007	·00006	·00006	·00005
10	·00005	·00004	·00004	·00003	·00003	·00003	·00002	·00002	·00002	·00002

For further values use the table at the foot of p.30;
e.g., $e^{-12 \cdot 5} = e^{-13 \cdot 1855} \times e^{1 \cdot 3155} = 10^{-6} \times 3 \cdot 727$.

Table 15. sinh x

x	·00	·01	·02	·03	·04	·05	·06	·07	·08	·09
0·0	0·0000	0·0100	0·0200	0·0300	0·0400	0·0500	0·0600	0·0701	0·0801	0·0901
0·1	0·1002	0·1102	0·1203	0·1304	0·1405	0·1506	0·1607	0·1708	0·1810	0·1911
0·2	0·2013	0·2115	0·2218	0·2320	0·2423	0·2526	0·2629	0·2733	0·2837	0·2941
0·3	0·3045	0·3150	0·3255	0·3360	0·3466	0·3572	0·3678	0·3785	0·3892	0·4000
0·4	0·4108	0·4216	0·4325	0·4434	0·4543	0·4653	0·4764	0·4875	0·4986	0·5098
0·5	0·5211	0·5324	0·5438	0·5552	0·5666	0·5782	0·5897	0·6014	0·6131	0·6248
0·6	0·6367	0·6485	0·6605	0·6725	0·6846	0·6967	0·7090	0·7213	0·7336	0·7461
0·7	0·7586	0·7712	0·7838	0·7966	0·8094	0·8223	0·8353	0·8484	0·8615	0·8748
0·8	0·8881	0·9015	0·9150	0·9286	0·9423	0·9561	0·9700	0·9840	0·9981	1·0122
0·9	1·0265	1·0409	1·0554	1·0700	1·0847	1·0995	1·1144	1·1294	1·1446	1·1598
1·0	1·1752	1·1907	1·2063	1·2220	1·2379	1·2539	1·2700	1·2862	1·3025	1·3190
1·1	1·3356	1·3524	1·3693	1·3863	1·4035	1·4208	1·4382	1·4558	1·4735	1·4914
1·2	1·5095	1·5276	1·5460	1·5645	1·5831	1·6019	1·6209	1·6400	1·6593	1·6788
1·3	1·6984	1·7182	1·7381	1·7583	1·7786	1·7991	1·8198	1·8406	1·8617	1·8829
1·4	1·9043	1·9259	1·9477	1·9697	1·9919	2·0143	2·0369	2·0597	2·0827	2·1059
1·5	2·1293	2·1529	2·1768	2·2008	2·2251	2·2496	2·2743	2·2993	2·3245	2·3499
1·6	2·3756	2·4015	2·4276	2·4540	2·4806	2·5075	2·5346	2·5620	2·5896	2·6175
1·7	2·6456	2·6740	2·7027	2·7317	2·7609	2·7904	2·8202	2·8503	2·8806	2·9112
1·8	2·9422	2·9734	3·0049	3·0367	3·0689	3·1013	3·1340	3·1671	3·2005	3·2341
1·9	3·2682	3·3025	3·3372	3·3722	3·4075	3·4432	3·4792	3·5156	3·5523	3·5894
2·0	3·6269	3·6647	3·7028	3·7414	3·7803	3·8196	3·8593	3·8993	3·9398	3·9806
2·1	4·0219	4·0635	4·1056	4·1480	4·1909	4·2342	4·2779	4·3220	4·3666	4·4116
2·2	4·4571	4·5030	4·5494	4·5962	4·6434	4·6912	4·7394	4·7880	4·8372	4·8868
2·3	4·9370	4·9876	5·0387	5·0903	5·1425	5·1951	5·2483	5·3020	5·3562	5·4109
2·4	5·4662	5·5221	5·5785	5·6354	5·6929	5·7510	5·8097	5·8689	5·9288	5·9892
2·5	6·0502	6·1118	6·1741	6·2369	6·3004	6·3645	6·4293	6·4946	6·5607	6·6274
2·6	6·6947	6·7628	6·8315	6·9008	6·9709	7·0417	7·1132	7·1854	7·2583	7·3319
2·7	7·4063	7·4814	7·5572	7·6338	7·7112	7·7894	7·8683	7·9480	8·0285	8·1098
2·8	8·1919	8·2749	8·3586	8·4432	8·5287	8·6150	8·7021	8·7902	8·8791	8·9689
2·9	9·0596	9·1512	9·2437	9·3371	9·4315	9·5268	9·6231	9·7203	9·8185	9·9177
3·0	10·018	10·119	10·221	10·325	10·429	10·534	10·640	10·748	10·856	10·966
3·1	11·076	11·188	11·301	11·415	11·530	11·647	11·764	11·883	12·003	12·124
3·2	12·246	12·369	12·494	12·620	12·747	12·876	13·006	13·137	13·269	13·403
3·3	13·538	13·674	13·812	13·951	14·092	14·234	14·377	14·522	14·668	14·816
3·4	14·965	15·116	15·268	15·422	15·577	15·734	15·893	16·053	16·214	16·378
3·5	16·543	16·709	16·877	17·047	17·219	17·392	17·567	17·744	17·923	18·103
3·6	18·285	18·470	18·655	18·843	19·033	19·224	19·418	19·613	19·811	20·010
3·7	20·211	20·415	20·620	20·828	21·037	21·249	21·463	21·679	21·897	22·117
3·8	22·339	22·564	22·791	23·020	23·252	23·486	23·722	23·961	24·202	24·445
3·9	24·691	24·939	25·190	25·444	25·700	25·958	26·219	26·483	26·749	27·018
4·0	27·290	27·564	27·842	28·122	28·404	28·690	28·979	29·270	29·564	29·862

For further values use $\sinh x = \frac{1}{2}(e^{x} - e^{-x})$.

Table 16. cosh x

x	·00	·01	·02	·03	·04	·05	·06	·07	·08	·09
0·0	1·0000	1·0001	1·0002	1·0005	1·0008	1·0013	1·0018	1·0025	1·0032	1·0041
0·1	1·0050	1·0061	1·0072	1·0085	1·0098	1·0113	1·0128	1·0145	1·0162	1·0181
0·2	1·0201	1·0221	1·0243	1·0266	1·0289	1·0314	1·0340	1·0367	1·0395	1·0423
0·3	1·0453	1·0484	1·0516	1·0549	1·0584	1·0619	1·0655	1·0692	1·0731	1·0770
0·4	1·0811	1·0852	1·0895	1·0939	1·0984	1·1030	1·1077	1·1125	1·1174	1·1225
0·5	1·1276	1·1329	1·1383	1·1438	1·1494	1·1551	1·1609	1·1669	1·1730	1·1792
0·6	1·1855	1·1919	1·1984	1·2051	1·2119	1·2188	1·2258	1·2330	1·2402	1·2476
0·7	1·2552	1·2628	1·2706	1·2785	1·2865	1·2947	1·3030	1·3114	1·3199	1·3286
0·8	1·3374	1·3464	1·3555	1·3647	1·3740	1·3835	1·3932	1·4029	1·4128	1·4229
0·9	1·4331	1·4434	1·4539	1·4645	1·4753	1·4862	1·4973	1·5085	1·5199	1·5314
1·0	1·5431	1·5549	1·5669	1·5790	1·5913	1·6038	1·6164	1·6292	1·6421	1·6552
1·1	1·6685	1·6820	1·6956	1·7093	1·7233	1·7374	1·7517	1·7662	1·7808	1·7957
1·2	1·8107	1·8258	1·8412	1·8568	1·8725	1·8884	1·9045	1·9208	1·9373	1·9540
1·3	1·9709	1·9880	2·0053	2·0228	2·0404	2·0583	2·0764	2·0947	2·1132	2·1320
1·4	2·1509	2·1700	2·1894	2·2090	2·2288	2·2488	2·2691	2·2896	2·3103	2·3312
1·5	2·3524	2·3738	2·3955	2·4174	2·4395	2·4619	2·4845	2·5073	2·5305	2·5538
1·6	2·5775	2·6013	2·6255	2·6499	2·6746	2·6995	2·7247	2·7502	2·7760	2·8020
1·7	2·8283	2·8549	2·8818	2·9090	2·9364	2·9642	2·9922	3·0206	3·0492	3·0782
1·8	3·1075	3·1370	3·1669	3·1971	3·2277	3·2585	3·2897	3·3212	3·3530	3·3852
1·9	3·4177	3·4506	3·4838	3·5173	3·5512	3·5855	3·6201	3·6551	3·6904	3·7261
2·0	3·7622	3·7987	3·8355	3·8727	3·9103	3·9483	3·9867	4·0255	4·0647	4·1043
2·1	4·1443	4·1847	4·2256	4·2668	4·3085	4·3507	4·3932	4·4362	4·4797	4·5236
2·2	4·5679	4·6127	4·6580	4·7037	4·7499	4·7966	4·8437	4·8914	4·9395	4·9881
2·3	5·0372	5·0868	5·1370	5·1876	5·2388	5·2905	5·3427	5·3954	5·4487	5·5026
2·4	5·5569	5·6119	5·6674	5·7235	5·7801	5·8373	5·8951	5·9535	6·0125	6·0721
2·5	6·1323	6·1931	6·2545	6·3166	6·3793	6·4426	6·5066	6·5712	6·6365	6·7024
2·6	6·7690	6·8363	6·9043	6·9729	7·0423	7·1123	7·1831	7·2546	7·3268	7·3998
2·7	7·4735	7·5479	7·6231	7·6990	7·7758	7·8533	7·9316	8·0106	8·0905	8·1712
2·8	8·2527	8·3351	8·4182	8·5022	8·5871	8·6728	8·7594	8·8469	8·9352	9·0244
2·9	9·1146	9·2056	9·2976	9·3905	9·4843	9·5791	9·6749	9·7716	9·8693	9·9680
3·0	10·068	10·168	10·270	10·373	10·477	10·581	10·687	10·794	10·902	11·011
3·1	11·122	11·233	11·345	11·459	11·574	11·690	11·807	11·925	12·044	12·165
3·2	12·287	12·410	12·534	12·660	12·786	12·915	13·044	13·175	13·307	13·440
3·3	13·575	13·711	13·848	13·987	14·127	14·269	14·412	14·557	14·702	14·850
3·4	14·999	15·149	15·301	15·455	15·610	15·766	15·924	16·084	16·245	16·408
3·5	16·573	16·739	16·907	17·077	17·248	17·421	17·596	17·772	17·951	18·131
3·6	18·313	18·497	18·682	18·870	19·059	19·250	19·444	19·639	19·836	20·035
3·7	20·236	20·439	20·644	20·852	21·061	21·272	21·486	21·702	21·919	22·140
3·8	22·362	22·586	22·813	23·042	23·274	23·507	23·743	23·982	24·222	24·466
3·9	24·711	24·960	25·210	25·463	25·719	25·977	26·238	26·502	26·768	27·037
4·0	27·308	27·583	27·860	28·139	28·422	28·707	28·996	29·287	29·581	29·878

For further values use $\cosh x = \frac{1}{2}(e^{x} + e^{-x})$.

Table 17. Degrees to Radians — Minutes to Radians

deg	radians	deg	radians	deg	radians	min	radians	min	radians
0°	0·0000	30°	0·5236	60°	1·0472	0′	0·0000	30′	0·0087
1	0·0175	31	0·5411	61	1·0647	1	0·0003	31	0·0090
2	0·0349	32	0·5585	62	1·0821	2	0·0006	32	0·0093
3	0·0524	33	0·5760	63	1·0996	3	0·0009	33	0·0096
4	0·0698	34	0·5934	64	1·1170	4	0·0012	34	0·0099
5	0·0873	35	0·6109	65	1·1345	5	0·0015	35	0·0102
6	0·1047	36	0·6283	66	1·1519	6	0·0017	36	0·0105
7	0·1222	37	0·6458	67	1·1694	7	0·0020	37	0·0108
8	0·1396	38	0·6632	68	1·1868	8	0·0023	38	0·0111
9	0·1571	39	0·6807	69	1·2043	9	0·0026	39	0·0113
10	0·1745	40	0·6981	70	1·2217	10	0·0029	40	0·0116
11	0·1920	41	0·7156	71	1·2392	11	0·0032	41	0·0119
12	0·2094	42	0·7330	72	1·2566	12	0·0035	42	0·0122
13	0·2269	43	0·7505	73	1·2741	13	0·0038	43	0·0125
14	0·2443	44	0·7679	74	1·2915	14	0·0041	44	0·0128
15	0·2618	45	0·7854	75	1·3090	15	0·0044	45	0·0131
16	0·2793	46	0·8029	76	1·3265	16	0·0047	46	0·0134
17	0·2967	47	0·8203	77	1·3439	17	0·0049	47	0·0137
18	0·3142	48	0·8378	78	1·3614	18	0·0052	48	0·0140
19	0·3316	49	0·8552	79	1·3788	19	0·0055	49	0·0143
20	0·3491	50	0·8727	80	1·3963	20	0·0058	50	0·0145
21	0·3665	51	0·8901	81	1·4137	21	0·0061	51	0·0148
22	0·3840	52	0·9076	82	1·4312	22	0·0064	52	0·0151
23	0·4014	53	0·9250	83	1·4486	23	0·0067	53	0·0154
24	0·4189	54	0·9425	84	1·4661	24	0·0070	54	0·0157
25	0·4363	55	0·9559	85	1·4835	25	0·0073	55	0·0160
26	0·4538	56	0·9774	86	1·5010	26	0·0076	56	0·0163
27	0·4712	57	0·9948	87	1·5184	27	0·0079	57	0·0166
28	0·4887	58	1·1023	88	1·5359	28	0·0081	58	0·0169
29	0·5061	59	1·0297	89	1·5533	29	0·0084	59	0·0172
30	0·5236	60	1·0472	90	1·5708	30	0·0087	60	0·0175

Table 18. Radians to degrees, and their trigonometric functions

rad	degrees	sin	cos	tan	rad	degrees	sin	cos	tan
0·01	0·57	0·0100	1·0000	0·0100	0·82	46·98	0·7311	0·6822	1·0717
0·02	1·15	0·0200	0·9998	0·0200	0·84	48·13	0·7446	0·6675	1·1156
0·03	1·72	0·0300	0·9996	0·0300	0·86	49·27	0·7578	0·6524	1·1616
0·04	2·29	0·0400	0·9992	0·0400	0·88	50·42	0·7707	0·6372	1·2097
0·05	2·86	0·0500	0·9988	0·0500	0·90	51·57	0·7833	0·6216	1·2602
0·06	3·44	0·0600	0·9982	0·0601	0·92	52·71	0·7956	0·6058	1·3133
0·07	4·01	0·0699	0·9976	0·0701	0·94	53·86	0·8076	0·5898	1·3692
0·08	4·58	0·0799	0·9968	0·0802	0·96	55·00	0·8192	0·5735	1·4284
0·09	5·16	0·0899	0·9960	0·0902	0·98	56·15	0·8305	0·5570	1·4910
0·10	5·73	0·0998	0·9950	0·1003	1·00	57·30	0·8415	0·5403	1·5574
0·12	6·88	0·1197	0·9928	0·1206	1·02	58·44	0·8521	0·5234	1·6281
0·14	8·02	0·1395	0·9902	0·1409	1·04	59·59	0·8624	0·5062	1·7036
0·16	9·17	0·1593	0·9872	0·1614	1·06	60·73	0·8724	0·4889	1·7844
0·18	10·31	0·1790	0·9838	0·1820	1·08	61·88	0·8820	0·4713	1·8712
0·20	11·46	0·1987	0·9801	0·2027	1·10	63·03	0·8912	0·4536	1·9648
0·22	12·61	0·2182	0·9759	0·2236	1·12	64·17	0·9001	0·4357	2·0660
0·24	13·75	0·2377	0·9713	0·2447	1·14	65·32	0·9086	0·4176	2·1759
0·26	14·90	0·2571	0·9664	0·2660	1·16	66·46	0·9168	0·3993	2·2958
0·28	16·04	0·2764	0·9611	0·2876	1·18	67·61	0·9246	0·3809	2·4273
0·30	17·19	0·2955	0·9553	0·3093	1·20	68·75	0·9320	0·3624	2·5722
0·32	18·33	0·3146	0·9492	0·3314	1·22	69·90	0·9391	0·3436	2·7328
0·34	19·48	0·3335	0·9428	0·3537	1·24	71·05	0·9458	0·3248	2·9119
0·36	20·63	0·3523	0·9359	0·3764	1·26	72·19	0·9521	0·3058	3·1133
0·38	21·77	0·3709	0·9287	0·3994	1·28	73·34	0·9580	0·2867	3·3413
0·40	22·92	0·3894	0·9211	0·4228	1·30	74·48	0·9636	0·2675	3·6021
0·42	24·06	0·4078	0·9131	0·4466	1·32	75·63	0·9687	0·2482	3·9033
0·44	25·21	0·4259	0·9048	0·4708	1·34	76·78	0·9735	0·2288	4·2556
0·46	26·36	0·4439	0·8961	0·4954	1·36	77·92	0·9779	0·2092	4·6734
0·48	27·50	0·4618	0·8870	0·5206	1·38	79·07	0·9819	0·1896	5·1774
0·50	28·65	0·4794	0·8776	0·5463	1·40	80·21	0·9854	0·1700	5·7979
0·52	29·79	0·4969	0·8678	0·5726	1·42	81·36	0·9887	0·1502	6·5811
0·54	30·94	0·5141	0·8577	0·5994	1·44	82·51	0·9915	0·1304	7·6018
0·56	32·09	0·5312	0·8473	0·6269	1·46	83·65	0·9939	0·1106	8·9886
0·58	33·23	0·5480	0·8365	0·6552	1·48	84·80	0·9959	0·0907	10·9834
0·60	34·38	0·5646	0·8253	0·6841	1·50	85·94	0·9975	0·0707	14·1014
0·62	35·52	0·5810	0·8139	0·7139	1·52	87·09	0·9987	0·0508	19·6695
0·64	36·67	0·5972	0·8021	0·7445	1·54	88·24	0·9995	0·0308	32·4611
0·66	37·82	0·6131	0·7900	0·7761	1·56	89·38	0·9999	0·0108	92·6205
0·68	38·96	0·6288	0·7776	0·8087					
0·70	40·11	0·6442	0·7648	0·8423					
0·72	41·25	0·6594	0·7518	0·8771					
0·74	42·40	0·6743	0·7385	0·9131					
0·76	43·54	0·6889	0·7248	0·9505					
0·78	44·69	0·7033	0·7109	0·9893					
0·80	45·84	0·7174	0·6967	1·0296					

Table 19. Prime numbers, 2–10271

2	199	467	769	1087	1429	1741	2089	2437	2791	3187	3541	3911	4271	4663
3	211	479	773	1091	1433	1747	2099	2441	2797	3191	3547	3917	4273	4673
5	223	487	787	1093	1439	1753	2111	2447	2801	3203	3557	3919	4283	4679
7	227	491	797	1097	1447	1759	2113	2459	2803	3209	3559	3923	4289	4691
11	229	499	809	1103	1451	1777	2129	2467	2819	3217	3571	3929	4297	4703
13	233	503	811	1109	1453	1783	2131	2473	2833	3221	3581	3931	4327	4721
17	239	509	821	1117	1459	1787	2137	2477	2837	3229	3583	3943	4337	4723
19	241	521	823	1123	1471	1789	2141	2503	2843	3251	3593	3947	4339	4729
23	251	523	827	1129	1481	1801	2143	2521	2851	3253	3607	3967	4349	4733
29	257	541	829	1151	1483	1811	2153	2531	2857	3257	3613	3989	4357	4751
31	263	547	839	1153	1487	1823	2161	2539	2861	3259	3617	4001	4363	4759
37	269	557	853	1163	1489	1831	2179	2543	2879	3271	3623	4003	4373	4783
41	271	563	857	1171	1493	1847	2203	2549	2887	3299	3631	4007	4391	4787
43	277	569	859	1181	1499	1861	2207	2551	2897	3301	3637	4013	4397	4789
47	281	571	863	1187	1511	1867	2213	2557	2903	3307	3643	4019	4409	4793
53	283	577	877	1193	1523	1871	2221	2579	2909	3313	3659	4021	4421	4799
59	293	587	881	1201	1531	1873	2237	2591	2917	3319	3671	4027	4423	4801
61	307	593	883	1213	1543	1877	2239	2593	2927	3323	3673	4049	4441	4813
67	311	599	887	1217	1549	1879	2243	2609	2939	3329	3677	4051	4447	4817
71	313	601	907	1223	1553	1889	2251	2617	2953	3331	3691	4057	4451	4831
73	317	607	911	1229	1559	1901	2267	2621	2957	3343	3697	4073	4457	4861
79	331	613	919	1231	1567	1907	2269	2633	2963	3347	3701	4079	4463	4871
83	337	617	929	1237	1571	1913	2273	2647	2969	3359	3709	4091	4481	4877
89	347	619	937	1249	1579	1931	2281	2657	2971	3361	3719	4093	4483	4889
97	349	631	941	1259	1583	1933	2287	2659	2999	3371	3727	4099	4493	4903
101	353	641	947	1277	1597	1949	2293	2663	3001	3373	3733	4111	4507	4909
103	359	643	953	1279	1601	1951	2297	2671	3011	3389	3739	4127	4513	4919
107	367	647	967	1283	1607	1973	2309	2677	3019	3391	3761	4129	4517	4931
109	373	653	971	1289	1609	1979	2311	2683	3023	3407	3767	4133	4519	4933
113	379	659	977	1291	1613	1987	2333	2687	3037	3413	3769	4139	4523	4937
127	383	661	983	1297	1619	1993	2339	2689	3041	3433	3779	4153	4547	4943
131	389	673	991	1301	1621	1997	2341	2693	3049	3449	3793	4157	4549	4951
137	397	677	997	1303	1627	1999	2347	2699	3061	3457	3797	4159	4561	4957
139	401	683	1009	1307	1637	2003	2351	2707	3067	3461	3803	4177	4567	4967
149	409	691	1013	1319	1657	2011	2357	2711	3079	3463	3821	4201	4583	4969
151	419	701	1019	1321	1663	2017	2371	2713	3083	3467	3823	4211	4591	4973
157	421	709	1021	1327	1667	2027	2377	2719	3089	3469	3833	4217	4597	4987
163	431	719	1031	1361	1669	2029	2381	2729	3109	3491	3847	4219	4603	4993
167	433	727	1033	1367	1693	2039	2383	2731	3119	3499	3851	4229	4621	4999
173	439	733	1039	1373	1697	2053	2389	2741	3121	3511	3853	4231	4637	5003
179	443	739	1049	1381	1699	2063	2393	2749	3137	3517	3863	4241	4639	5009
181	449	743	1051	1399	1709	2069	2399	2753	3163	3527	3877	4243	4643	5011
191	457	751	1061	1409	1721	2081	2411	2767	3167	3529	3881	4253	4649	5021
193	461	757	1063	1423	1723	2083	2417	2777	3169	3533	3889	4259	4651	5023
197	463	761	1069	1427	1733	2087	2423	2789	3181	3539	3907	4261	4657	5039

Prime numbers, 2–10271

5051	5449	5839	6229	6637	7001	7477	7841	8263	8681	9059	9461	9857
5059	5471	5843	6247	6653	7013	7481	7853	8269	8689	9067	9463	9859
5077	5477	5849	6257	6659	7019	7487	7867	8273	8693	9091	9467	9871
5081	5479	5851	6263	6661	7027	7489	7873	8287	8699	9103	9473	9883
5087	5483	5857	6269	6673	7039	7499	7877	8291	8707	9109	9479	9887
5099	5501	5861	6271	6679	7043	7507	7879	8293	8713	9127	9491	9901
5101	5503	5867	6277	6689	7057	7517	7883	8297	8719	9133	9497	9907
5107	5507	5869	6287	6691	7069	7523	7901	8311	8731	9137	9511	9923
5113	5519	5879	6299	6701	7079	7529	7907	8317	8737	9151	9521	9929
5119	5521	5881	6301	6703	7103	7537	7919	8329	8741	9157	9533	9931
5147	5527	5897	6311	6709	7109	7541	7927	8353	8747	9161	9539	9941
5153	5531	5903	6317	6719	7121	7547	7933	8363	8753	9173	9547	9949
5167	5557	5923	6323	6733	7127	7549	7937	8369	8761	9181	9551	9967
5171	5563	5927	6329	6737	7129	7559	7949	8377	8779	9187	9587	9973
5179	5569	5939	6337	6761	7151	7561	7951	8387	8783	9199	9601	10007
5189	5573	5953	6343	6763	7159	7573	7963	8389	8803	9203	9613	10009
5197	5581	5981	6353	6779	7177	7577	7993	8419	8807	9209	9619	10037
5209	5591	5987	6359	6781	7187	7583	8009	8423	8819	9221	9623	10039
5227	5623	6007	6361	6791	7193	7589	8011	8429	8821	9227	9629	10061
5231	5639	6011	6367	6793	7207	7591	8017	8431	8831	9239	9631	10067
5233	5641	6029	6373	6803	7211	7603	8039	8443	8837	9241	9643	10069
5237	5647	6037	6379	6823	7213	7607	8053	8447	8839	9257	9649	10079
5261	5651	6043	6389	6827	7219	7621	8059	8461	8849	9277	9661	10091
5273	5653	6047	6397	6829	7229	7639	8069	8467	8861	9281	9677	10093
5279	5657	6053	6421	6833	7237	7643	8081	8501	8863	9283	9679	10099
5281	5659	6067	6427	6841	7243	7649	8087	8513	8867	9293	9689	10103
5297	5669	6073	6449	6857	7247	7669	8089	8521	8887	9311	9697	10111
5303	5683	6079	6451	6863	7253	7673	8093	8527	8893	9319	9719	10133
5309	5689	6089	6469	6869	7283	7681	8101	8537	8923	9323	9721	10139
5323	5693	6091	6473	6871	7297	7687	8111	8539	8929	9337	9733	10141
5333	5701	6101	6481	6883	7307	7691	8117	8543	8933	9341	9739	10151
5347	5711	6113	6491	6899	7309	7699	8123	8563	8941	9343	9743	10159
5351	5717	6121	6521	6907	7321	7703	8147	8573	8951	9349	9749	10163
5381	5737	6131	6529	6911	7331	7717	8161	8581	8963	9371	9767	10169
5387	5741	6133	6547	6917	7333	7723	8167	8597	8969	9377	9769	10177
5393	5743	6143	6551	6947	7349	7727	8171	8599	8971	9391	9781	10181
5399	5749	6151	6553	6949	7351	7741	8179	8609	8999	9397	9787	10193
5407	5779	6163	6563	6959	7369	7753	8191	8623	9001	9403	9791	10211
5413	5783	6173	6569	6961	7393	7757	8209	8627	9007	9413	9803	10223
5417	5791	6197	6571	6967	7411	7759	8219	8629	9011	9419	9811	10243
5419	5801	6199	6577	6971	7417	7789	8221	8641	9013	9421	9817	10247
5431	5807	6203	6581	6977	7433	7793	8231	8647	9029	9431	9829	10253
5437	5813	6211	6599	6983	7451	7817	8233	8663	9041	9433	9833	10259
5441	5821	6217	6607	6991	7457	7823	8237	8669	9043	9437	9839	10267
5443	5827	6221	6619	6997	7459	7829	8243	8677	9049	9439	9851	10271

Table 20. Logarithms of factorials

x	$\log_{10} x!$	x	$\log_{10} x!$	x	$\log_{10} x!$	x	$\log_{10} x!$
1	0·0000	26	26·6056	51	66·1906	76	111·2754
2	0·3010	27	28·0370	52	67·9066	77	113·1619
3	0·7782	28	29·4841	53	69·6309	78	115·0540
4	1·3802	29	30·9465	54	71·3633	79	116·9516
5	2·0792	30	32·4237	55	73·1037	80	118·8547
6	2·8573	31	33·9150	56	74·8519	81	120·7632
7	3·7024	32	35·4202	57	76·6077	82	122·6770
8	4·6055	33	36·9387	58	78·3712	83	124·5961
9	5·5598	34	38·4702	59	80·1420	84	126·5204
10	6·5598	35	40·0142	60	81·9202	85	128·4498
11	7·6012	36	41·5705	61	83·7055	86	130·3843
12	8·6803	37	43·1387	62	85·4979	87	132·3238
13	9·7943	38	44·7185	63	87·2972	88	134·2683
14	10·9404	39	46·3096	64	89·1034	89	136·2177
15	12·1165	40	47·9116	65	90·9163	90	138·1719
16	13·3206	41	49·5244	66	92·7359	91	140·1310
17	14·5511	42	51·1477	67	94·5619	92	142·0948
18	15·8063	43	52·7811	68	96·3945	93	144·0632
19	17·0851	44	54·4246	69	98·2333	94	146·0364
20	18·3861	45	56·0778	70	100·0784	95	148·0141
21	19·7083	46	57·7406	71	101·9297	96	149·9964
22	21·0508	47	59·4127	72	103·7870	97	151·9831
23	22·4125	48	61·0939	73	105·6503	98	153·9744
24	23·7927	49	62·7841	74	107·5196	99	155·9700
25	25·1906	50	64·4831	75	109·3946	100	157·9700

For further values use:

$\log_{10} x! \approx 0{\cdot}3991 + (x+\tfrac{1}{2}) \log_{10} x - 0{\cdot}434295\, x$.

Table 21. Binomial coefficients $\binom{n}{r}$

n \ r	0	1	2	3	4	5	6	7	8	9	10
1	1	1									
2	1	2	1								
3	1	3	3	1							
4	1	4	6	4	1						
5	1	5	10	10	5	1					
6	1	6	15	20	15	6	1				
7	1	7	21	35	35	21	7	1			
8	1	8	28	56	70	56	28	8	1		
9	1	9	36	84	126	126	84	36	9	1	
10	1	10	45	120	210	252	210	120	45	10	1
11	1	11	55	165	330	462	462	330	165	55	11
12	1	12	66	220	495	792	924	792	495	220	66
13	1	13	78	286	715	1287	1716	1716	1287	715	286
14	1	14	91	364	1001	2002	3003	3432	3003	2002	1001
15	1	15	105	455	1365	3003	5005	6435	6435	5005	3003
16	1	16	120	560	1820	4368	8008	11440	12870	11440	8008
17	1	17	136	680	2380	6188	12376	19448	24310	24310	19448
18	1	18	153	816	3060	8568	18564	31824	43758	48620	43758
19	1	19	171	969	3876	11628	27132	50388	75582	92378	92378
20	1	20	190	1140	4845	15504	38760	77520	125970	167960	184756

Table 20 may be used for further values;

e.g., $\log \binom{52}{13} = \log 52! - \log 13! - \log 39!$
$= 67{\cdot}9066 - 9{\cdot}7943 - 46{\cdot}3096 = 11{\cdot}8027;$

$\binom{52}{13} = 6{\cdot}349 \times 10^{11}.$

Table 22. The sign test: cumulative binomial probabilities for distributions with $p = \frac{1}{2}$

n \ r	0	1	2	3	4	5	6	7	8	9	10	11	12
5	0·0313	0·1875	0·5000										
6	0·0156	0·1094	0·3438	0·6563									
7	0·0078	0·0625	0·2266	0·5000									
8	0·0039	0·0352	0·1445	0·3633	0·6367								
9	0·0020	0·0195	0·0898	0·2539	0·5000								
10	0·0010	0·0107	0·0547	0·1719	0·3770	0·6230							
11	0·0005	0·0059	0·0327	0·1133	0·2744	0·5000							
12	0·0002	0·0032	0·0193	0·0730	0·1938	0·3872	0·6128						
13	0·0001	0·0017	0·0112	0·0461	0·1334	0·2905	0·5000						
14	0·0001	0·0009	0·0065	0·0287	0·0898	0·2120	0·3953	0·6047					
15		0·0005	0·0037	0·0176	0·0592	0·1509	0·3036	0·5000					
16		0·0003	0·0021	0·0106	0·0384	0·1051	0·2272	0·4018	0·5982				
17		0·0001	0·0012	0·0064	0·0245	0·0717	0·1662	0·3145	0·5000				
18		0·0001	0·0007	0·0038	0·0154	0·0481	0·1189	0·2403	0·4073	0·5927			
19			0·0004	0·0022	0·0096	0·0318	0·0835	0·1796	0·3238	0·5000			
20			0·0002	0·0013	0·0059	0·0207	0·0577	0·1316	0·2517	0·4119	0·5881		
21			0·0001	0·0007	0·0036	0·0133	0·0392	0·0946	0·1917	0·3318	0·5000		
22			0·0001	0·0004	0·0022	0·0085	0·0262	0·0669	0·1431	0·2617	0·4159	0·5841	
23				0·0002	0·0013	0·0053	0·0173	0·0466	0·1050	0·2024	0·3388	0·5000	
24				0·0001	0·0008	0·0033	0·0113	0·0320	0·0758	0·1537	0·2706	0·4194	0·5806
25				0·0001	0·0005	0·0020	0·0073	0·0216	0·0539	0·1148	0·2122	0·3450	0·5000

The tabulated value is the probability that $X \leqslant r$, where X has a binomial distribution with parameters n and p. The table is limited to distributions with $p = \frac{1}{2}$ and is suitable for use with the sign test.

Table 23. Cumulative Poisson probabilities

r \ λ	0·1	0·2	0·3	0·4	0·5	0·6	0·7	0·8	0·9	1·0	1·5	2·0	2·5
0	0·905	0·819	0·741	0·670	0·607	0·549	0·497	0·449	0·407	0·368	0·223	0·135	0·082
1	0·995	0·982	0·963	0·938	0·910	0·878	0·844	0·809	0·772	0·736	0·558	0·406	0·287
2	1·000	0·999	0·996	0·992	0·986	0·977	0·966	0·953	0·937	0·920	0·809	0·677	0·544
3		1·000	1·000	0·999	0·998	0·997	0·994	0·991	0·987	0·981	0·934	0·857	0·758
4				1·000	1·000	1·000	0·999	0·999	0·998	0·996	0·981	0·947	0·891
5							1·000	1·000	1·000	0·999	0·996	0·983	0·958
6										1·000	0·999	0·995	0·986
7											1·000	0·999	0·996
8												1·000	0·999
9													1·000

r \ λ	3·0	3·5	4·0	4·5	5·0	5·5	6·0	6·5	7·0	7·5	8·0	9·0	10·0
0	0·050	0·030	0·018	0·011	0·007	0·004	0·002	0·002	0·001	0·001	0·000	0·000	0·000
1	0·199	0·136	0·092	0·061	0·040	0·027	0·017	0·011	0·007	0·005	0·003	0·001	0·000
2	0·423	0·321	0·238	0·174	0·125	0·088	0·062	0·043	0·030	0·020	0·014	0·006	0·003
3	0·647	0·537	0·433	0·342	0·265	0·202	0·151	0·112	0·082	0·059	0·042	0·021	0·010
4	0·815	0·725	0·629	0·532	0·440	0·358	0·285	0·224	0·173	0·132	0·100	0·055	0·029
5	0·916	0·858	0·785	0·703	0·616	0·529	0·446	0·369	0·301	0·241	0·191	0·116	0·067
6	0·966	0·935	0·889	0·831	0·762	0·686	0·606	0·527	0·450	0·378	0·313	0·207	0·130
7	0·988	0·973	0·949	0·913	0·867	0·809	0·744	0·673	0·599	0·525	0·453	0·324	0·220
8	0·996	0·990	0·979	0·960	0·932	0·894	0·847	0·792	0·729	0·662	0·593	0·456	0·333
9	0·999	0·997	0·992	0·983	0·968	0·946	0·916	0·877	0·830	0·776	0·717	0·587	0·458
10	1·000	0·999	0·997	0·993	0·986	0·975	0·957	0·933	0·901	0·862	0·816	0·706	0·583
11		1·000	0·999	0·998	0·995	0·989	0·980	0·966	0·947	0·921	0·888	0·803	0·697
12			1·000	0·999	0·998	0·996	0·991	0·984	0·973	0·957	0·936	0·876	0·792
13				1·000	0·999	0·998	0·996	0·993	0·987	0·978	0·966	0·926	0·864
14					1·000	0·999	0·999	0·997	0·994	0·990	0·983	0·959	0·917
15						1·000	0·999	0·999	0·998	0·995	0·992	0·978	0·951
16							1·000	1·000	0·999	0·998	0·996	0·989	0·973
17									1·000	0·999	0·998	0·995	0·986
18										1·000	0·999	0·998	0·993
19											1·000	0·999	0·997
20												1·000	0·998
21													0·999
22													1·000

The tabulated value is the probability that $X \leqslant r$, where X has a Poisson distribution with mean λ. A tabulated value of 1·000 represents a probability greater than 0·9995.

Table 24. The normal distribution function $\Phi(x)$

x	·00	·01	·02	·03	·04	·05	·06	·07	·08	·09
0·0	0·5000	0·5040	0·5080	0·5120	0·5160	0·5199	0·5239	0·5279	0·5319	0·5359
0·1	0·5398	0·5438	0·5478	0·5517	0·5557	0·5596	0·5636	0·5675	0·5714	0·5753
0·2	0·5793	0·5832	0·5871	0·5910	0·5948	0·5987	0·6026	0·6064	0·6103	0·6141
0·3	0·6179	0·6217	0·6255	0·6293	0·6331	0·6368	0·6406	0·6443	0·6480	0·6517
0·4	0·6554	0·6591	0·6628	0·6664	0·6700	0·6736	0·6772	0·6808	0·6844	0·6879
0·5	0·6915	0·6950	0·6985	0·7019	0·7054	0·7088	0·7123	0·7157	0·7190	0·7224
0·6	0·7257	0·7291	0·7324	0·7357	0·7389	0·7422	0·7454	0·7486	0·7517	0·7549
0·7	0·7580	0·7611	0·7642	0·7673	0·7703	0·7734	0·7764	0·7793	0·7823	0·7852
0·8	0·7881	0·7910	0·7939	0·7967	0·7995	0·8023	0·8051	0·8078	0·8106	0·8133
0·9	0·8159	0·8186	0·8212	0·8238	0·8264	0·8289	0·8315	0·8340	0·8365	0·8389
1·0	0·8413	0·8438	0·8461	0·8485	0·8508	0·8531	0·8554	0·8577	0·8599	0·8621
1·1	0·8643	0·8665	0·8686	0·8708	0·8729	0·8749	0·8770	0·8790	0·8810	0·8830
1·2	0·8849	0·8869	0·8888	0·8906	0·8925	0·8943	0·8962	0·8980	0·8997	0·9015
1·3	0·9032	0·9049	0·9066	0·9082	0·9099	0·9115	0·9131	0·9147	0·9162	0·9177
1·4	0·9192	0·9207	0·9222	0·9236	0·9251	0·9265	0·9279	0·9292	0·9306	0·9319
1·5	0·9332	0·9345	0·9357	0·9370	0·9382	0·9394	0·9406	0·9418	0·9429	0·9441
1·6	0·9452	0·9463	0·9474	0·9484	0·9495	0·9505	0·9515	0·9525	0·9535	0·9545
1·7	0·9554	0·9564	0·9573	0·9582	0·9591	0·9599	0·9608	0·9616	0·9625	0·9633
1·8	0·9641	0·9648	0·9656	0·9664	0·9671	0·9678	0·9686	0·9693	0·9699	0·9706
1·9	0·9713	0·9719	0·9726	0·9732	0·9738	0·9744	0·9750	0·9756	0·9761	0·9767
2·0	0·9772	0·9778	0·9783	0·9788	0·9793	0·9798	0·9803	0·9808	0·9812	0·9817
2·1	0·9821	0·9826	0·9830	0·9834	0·9838	0·9842	0·9846	0·9850	0·9854	0·9857
2·2	0·9861	0·9864	0·9868	0·9871	0·9875	0·9878	0·9881	0·9884	0·9887	0·9890
2·3	0·9893	0·9896	0·9898	0·9901	0·9904	0·9906	0·9909	0·9911	0·9913	0·9916
2·4	0·9918	0·9920	0·9922	0·9924	0·9927	0·9929	0·9930	0·9932	0·9934	0·9936
2·5	0·9938	0·9940	0·9941	0·9943	0·9945	0·9946	0·9948	0·9949	0·9951	0·9952
2·6	0·9953	0·9955	0·9956	0·9957	0·9959	0·9960	0·9961	0·9962	0·9963	0·9964
2·7	0·9965	0·9966	0·9967	0·9968	0·9969	0·9970	0·9971	0·9972	0·9973	0·9974
2·8	0·9974	0·9975	0·9976	0·9977	0·9977	0·9978	0·9979	0·9979	0·9980	0·9981
2·9	0·9981	0·9982	0·9982	0·9983	0·9984	0·9984	0·9985	0·9985	0·9986	0·9986
3·0	0·9986	0·9987	0·9987	0·9988	0·9988	0·9989	0·9989	0·9989	0·9990	0·9990
3·1	0·9990	0·9991	0·9991	0·9991	0·9992	0·9992	0·9992	0·9992	0·9993	0·9993
3·2	0·9993	0·9993	0·9994	0·9994	0·9994	0·9994	0·9994	0·9995	0·9995	0·9995
3·3	0·9995	0·9995	0·9995	0·9996	0·9996	0·9996	0·9996	0·9996	0·9996	0·9996
3·4	0·9997	0·9997	0·9997	0·9997	0·9997	0·9997	0·9997	0·9997	0·9997	0·9998
3·5	0·9998	0·9998	0·9998	0·9998	0·9998	0·9998	0·9998	0·9998	0·9998	0·9998
3·6	0·9998	0·9998	0·9998	0·9999	0·9999	0·9999	0·9999	0·9999	0·9999	0·9999

The tabulated value $\Phi(x)$ is the probability that a random variable, normally distributed with zero mean and unit variance, will be less than x.

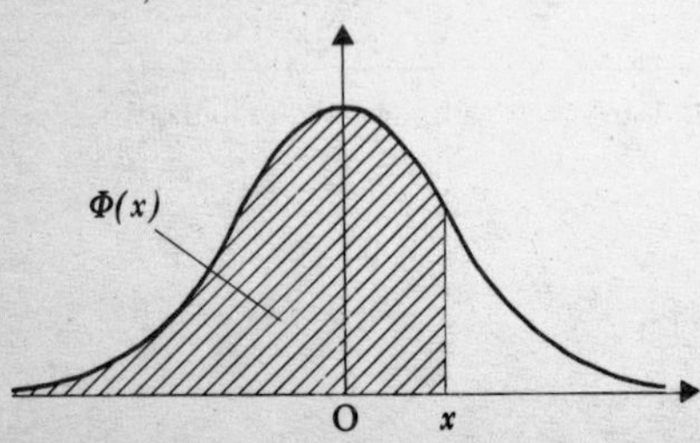

Table 25. The normal frequency function $\phi(x)$

x	·0	·1	·2	·3	·4	·5	·6	·7	·8	·9
0	0·3989	0·3970	0·3910	0·3814	0·3683	0·3521	0·3332	0·3123	0·2897	0·2661
1	0·2420	0·2179	0·1942	0·1714	0·1497	0·1295	0·1109	0·0940	0·0790	0·0656
2	0·0540	0·0440	0·0355	0·0283	0·0224	0·0175	0·0136	0·0104	0·0079	0·0060
3	0·0044	0·0033	0·0024	0·0017	0·0012	0·0009	0·0006	0·0004	0·0003	0·0002
4	0·0001									

$$\phi(x) = \frac{1}{\sqrt{2\pi}} e^{-\frac{1}{2}x^2}$$

$\phi(x)$ is the ordinate of the normal frequency curve.

Table 26. The upper percentage points of the normal distribution

P	·05	·025	·01	·005	·001	·0005
x_P	1·6449	1·9600	2·3263	2·5758	3·0902	3·2905

If a random variable is normally distributed with zero mean and unit variance, the probability that it takes a value exceeding x_P is P.

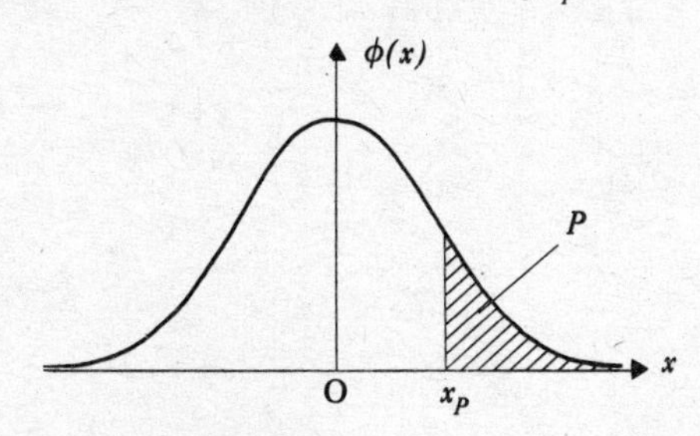

Table 27. Critical values of r, the product-moment correlation coefficient

ν \ P	·05	·025	·01	·005	·0005
1	0·98769	0·99692	0·999507	0·999877	0·9999988
2	0·90000	0·95000	0·98000	0·990000	0·99900
3	0·8054	0·8783	0·93433	0·95873	0·99116
4	0·7293	0·8114	0·8822	0·91720	0·97406
5	0·6694	0·7545	0·8329	0·8745	0·95074
6	0·6215	0·7067	0·7887	0·8343	0·92493
7	0·5822	0·6664	0·7498	0·7977	0·8982
8	0·5494	0·6319	0·7155	0·7646	0·8721
9	0·5214	0·6021	0·6851	0·7348	0·8471
10	0·4973	0·5760	0·6581	0·7079	0·8233
11	0·4762	0·5529	0·6339	0·6835	0·8010
12	0·4575	0·5324	0·6120	0·6614	0·7800
13	0·4409	0·5139	0·5923	0·6411	0·7603
14	0·4259	0·4973	0·5742	0·6226	0·7420
15	0·4124	0·4821	0·5577	0·6055	0·7246
16	0·4000	0·4683	0·5425	0·5897	0·7084
17	0·3887	0·4555	0·5285	0·5751	0·6932
18	0·3783	0·4438	0·5155	0·5614	0·6787
19	0·3687	0·4329	0·5034	0·5487	0·6652
20	0·3598	0·4227	0·4921	0·5368	0·6524
25	0·3233	0·3809	0·4451	0·4869	0·5974
30	0·2960	0·3494	0·4093	0·4487	0·5541
35	0·2746	0·3246	0·3810	0·4182	0·5189
40	0·2573	0·3044	0·3578	0·3932	0·4896
45	0·2428	0·2875	0·3384	0·3721	0·4648
50	0·2306	0·2732	0·3218	0·3541	0·4433
60	0·2108	0·2500	0·2948	0·3248	0·4078
70	0·1954	0·2319	0·2737	0·3017	0·3799
80	0·1829	0·2172	0·2565	0·2830	0·3568
90	0·1726	0·2050	0·2422	0·2673	0·3375
100	0·1638	0·1946	0·2301	0·2540	0·3211

For a total correlation, ν is 2 less than the number of pairs in the sample; for a partial correlation, the number of eliminated variates should also be subtracted.

The tabulated value r_P is such that, if there is no correlation in the parent population, the probability that $r \geqslant r_P$ is P.

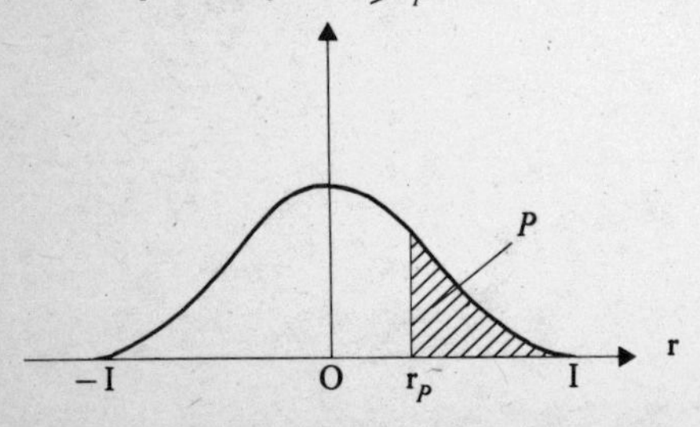

Table 28. Transformation of r $\tanh^{-1} r$

r	·00	·01	·02	·03	·04	·05	·06	·07	·08	·09
0·0	0·000	0·010	0·020	0·030	0·040	0·050	0·060	0·070	0·080	0·090
0·1	0·100	0·110	0·121	0·131	0·141	0·151	0·161	0·172	0·182	0·192
0·2	0·203	0·213	0·224	0·234	0·245	0·255	0·266	0·277	0·288	0·299
0·3	0·310	0·321	0·332	0·343	0·354	0·365	0·377	0·388	0·400	0·412
0·4	0·424	0·436	0·448	0·460	0·472	0·485	0·497	0·510	0·523	0·536
0·5	0·549	0·563	0·576	0·590	0·604	0·618	0·633	0·648	0·662	0·678
0·6	0·693	0·709	0·725	0·741	0·758	0·775	0·793	0·811	0·829	0·848
0·7	0·867	0·887	0·908	0·929	0·950	0·973	0·996	1·020	1·045	1·071
0·8	1·099	1·127	1·157	1·188	1·221	1·256	1·293	1·333	1·376	1·422

r	·000	·001	·002	·003	·004	·005	·006	·007	·008	·009
·90	1·472	1·478	1·483	1·488	1·494	1·499	1·505	1·510	1·516	1·522
·91	1·528	1·533	1·539	1·545	1·551	1·557	1·564	1·570	1·576	1·583
·92	1·589	1·596	1·602	1·609	1·616	1·623	1·630	1·637	1·644	1·651
·93	1·658	1·666	1·673	1·681	1·689	1·697	1·705	1·713	1·721	1·730
·94	1·738	1·747	1·756	1·764	1·774	1·783	1·792	1·802	1·812	1·822
·95	1·832	1·842	1·853	1·863	1·874	1·886	1·897	1·909	1·921	1·933
·96	1·946	1·959	1·972	1·986	2·000	2·014	2·029	2·044	2·060	2·076
·97	2·092	2·110	2·127	2·146	2·165	2·185	2·205	2·227	2·249	2·273
·98	2·298	2·323	2·351	2·380	2·410	2·443	2·477	2·515	2·555	2·599
·99	2·647	2·700	2·759	2·826	2·903	2·994	3·106	3·250	3·453	3·800

The function tabulated is $z = \tanh^{-1} r = \frac{1}{2}\ln\frac{1+r}{1-r}$.

For samples with large ν drawn from a population whose correlation coefficient is ρ, z is approximately normally distributed with mean $\tanh^{-1}\rho + \frac{\rho}{2(\nu+1)}$ and variance $\frac{1}{\nu-1}$.

Table 29. Critical values of r_S, the Spearman rank correlation coefficient

n \ P	·05	·01
4	1·000	
5	0·900	1·000
6	0·829	0·943
7	0·714	0·893
8	0·643	0·833
9	0·600	0·783
10	0·564	0·733
12	0·503	0·678
14	0·456	0·645
16	0·425	0·601
18	0·399	0·564
20	0·377	0·534
22	0·359	0·508
24	0·343	0·485
26	0·329	0·465
28	0·317	0·448
30	0·305	0·432

Spearman's r_S is calculated from a sample of n pairs drawn from a population of pairs. The tabulated value r_P is such that, if there is no correlation in the parent population, the probability that $r_S \geqslant r_P$ is *P*.

For values of n greater than 20, r_S is approximately normally distributed with mean zero and variance $\frac{1}{n-1}$.

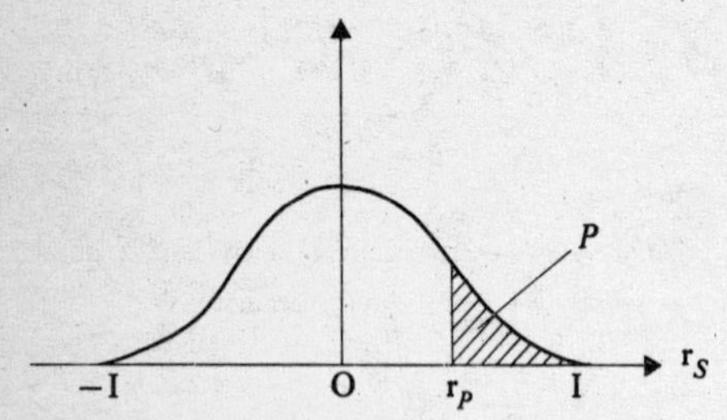

Table 30. The Kendall rank correlation coefficient: probability that S (for τ) attains or exceeds a specified value

S \ n	4	5	8	9
0	0·625	0·592	0·548	0·540
2	0·375	0·408	0·452	0·460
4	0·167	0·242	0·360	0·381
6	0·042	0·117	0·274	0·306
8		0·042	0·199	0·238
10		0·0083	0·138	0·179
12			0·089	0·130
14			0·054	0·090
16			0·031	0·060
18			0·016	0·038
20			0·0071	0·022
22			0·0028	0·012
24			0·00087	0·0063
26			0·00019	0·0029
28			0·000025	0·0012
30				0·00043
32				0·00012
34				0·000025
36				0·0000028

S \ n	6	7	10
1	0·500	0·500	0·500
3	0·360	0·386	0·431
5	0·235	0·281	0·364
7	0·136	0·191	0·300
9	0·068	0·119	0·242
11	0·028	0·068	0·190
13	0·0083	0·035	0·146
15	0·0014	0·015	0·108
17		0·0054	0·078
19		0·0014	0·054
21		0·00020	0·036
23			0·023
25			0·014
27			0·0083
29			0·0046
31			0·0023
33			0·0011
35			0·00047
37			0·00018
39			0·000058
41			0·000015
43			0·0000028
45			0·00000028

$$\tau = \frac{S}{\frac{1}{2}n\,(n-1)}$$

S (for τ) is calculated from a sample of n pairs drawn from a population of pairs. The table gives the probability that S attains or exceeds a specified value when there is no correlation in the parent population.

For $n \geqslant 10$, S is approximately normally distributed with mean zero and variance $\frac{1}{18}n(n-1)(2n+5)$.

Table 31. Upper percentage points of the t-distribution

ν \ P	·10	·05	·025	·01	·005	·001
1	3·078	6·314	12·706	31·821	63·657	318·310
2	1·886	2·920	4·303	6·965	9·925	22·327
3	1·638	2·353	3·182	4·541	5·841	10·213
4	1·533	2·132	2·776	3·747	4·604	7·173
5	1·476	2·015	2·571	3·365	4·032	5·893
6	1·440	1·943	2·447	3·143	3·707	5·208
7	1·415	1·895	2·365	2·998	3·499	4·785
8	1·397	1·860	2·306	2·896	3·355	4·501
9	1·383	1·833	2·262	2·821	3·250	4·297
10	1·372	1·812	2·228	2·764	3·169	4·144
11	1·363	1·796	2·201	2·718	3·106	4·025
12	1·356	1·782	2·179	2·681	3·055	3·930
13	1·350	1·771	2·160	2·650	3·012	3·852
14	1·345	1·761	2·145	2·624	2·977	3·787
15	1·341	1·753	2·131	2·602	2·947	3·733
16	1·337	1·746	2·120	2·583	2·921	3·686
17	1·333	1·740	2·110	2·567	2·898	3·646
18	1·330	1·734	2·101	2·552	2·878	3·610
19	1·328	1·729	2·093	2·539	2·861	3·579
20	1·325	1·725	2·086	2·528	2·845	3·552
21	1·323	1·721	2·080	2·518	2·831	3·527
22	1·321	1·717	2·074	2·508	2·819	3·505
23	1·319	1·714	2·069	2·500	2·807	3·485
24	1·318	1·711	2·064	2·492	2·797	3·467
25	1·316	1·708	2·060	2·485	2·787	3·450
26	1·315	1·706	2·056	2·479	2·779	3·435
27	1·314	1·703	2·052	2·473	2·771	3·421
28	1·313	1·701	2·048	2·467	2·763	3·408
29	1·311	1·699	2·045	2·462	2·756	3·396
30	1·310	1·697	2·042	2·457	2·750	3·385
40	1·303	1·684	2·021	2·423	2·704	3·307
60	1·296	1·671	2·000	2·390	2·660	3·232
120	1·289	1·658	1·980	2·358	2·617	3·160
∞	1·282	1·645	1·960	2·326	2·576	3·090

The tabulated value t_P is such that, if the random variable X has the t-distribution with ν degrees of freedom, the probability that $X \geqslant t_P$ is P.

For $\nu > 30$, interpolation ν-wise should be linear in $\frac{120}{\nu}$ (see p.7).

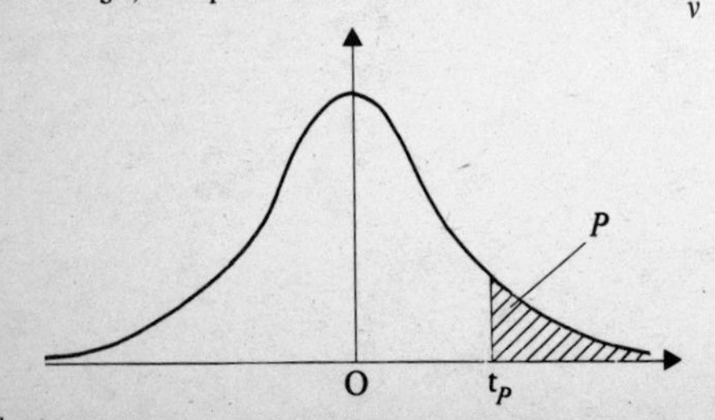

Table 32. Upper percentage points of the χ^2-distribution

ν \ P	·99	·95	·50	·20	·10	·05	·025	·01
1	0·0002	0·0039	0·45	1·64	2·71	3·84	5·02	6·63
2	0·020	0·103	1·39	3·22	4·61	5·99	7·38	9·21
3	0·115	0·352	2·37	4·64	6·25	7·81	9·35	11·34
4	0·30	0·71	3·36	5·99	7·78	9·49	11·14	13·28
5	0·55	1·15	4·35	7·29	9·24	11·07	12·83	15·09
6	0·87	1·64	5·35	8·56	10·64	12·59	14·45	16·81
7	1·24	2·17	6·35	9·80	12·02	14·07	16·01	18·48
8	1·65	2·73	7·34	11·03	13·36	15·51	17·53	20·09
9	2·09	3·33	8·34	12·24	14·68	16·92	19·02	21·67
10	2·56	3·94	9·34	13·44	15·99	18·31	20·48	23·21
11	3·05	4·57	10·34	14·63	17·28	19·68	21·92	24·72
12	3·57	5·23	11·34	15·81	18·55	21·03	23·34	26·22
13	4·11	5·89	12·34	16·98	19·81	22·36	24·74	27·69
14	4·66	6·57	13·34	18·15	21·06	23·68	26·12	29·14
15	5·23	7·26	14·34	19·31	22·31	25·00	27·49	30·58
16	5·81	7·96	15·34	20·47	23·54	26·30	28·85	32·00
17	6·41	8·67	16·34	21·61	24·77	27·59	30·19	33·41
18	7·02	9·39	17·34	22·76	25·99	28·87	31·53	34·81
19	7·63	10·12	18·34	23·90	27·20	30·14	32·85	36·19
20	8·26	10·85	19·34	25·04	28·41	31·41	34·17	37·57
21	8·90	11·59	20·34	26·17	29·62	32·67	35·48	38·93
22	9·54	12·34	21·34	27·30	30·81	33·92	36·78	40·29
23	10·20	13·09	22·34	28·43	32·01	35·17	38·08	41·64
24	10·86	13·85	23·34	29·55	33·20	36·42	39·36	42·98
25	11·52	14·61	24·34	30·68	34·38	37·65	40·65	44·31
26	12·20	15·38	25·34	31·79	35·56	38·89	41·92	45·64
27	12·88	16·15	26·34	32·91	36·74	40·11	43·19	46·96
28	13·57	16·93	27·34	34·03	37·92	41·34	44·46	48·28
29	14·26	17·71	28·34	35·14	39·09	42·56	45·72	49·59
30	14·95	18·49	29·34	36·25	40·26	43·77	46·98	50·89
40	22·16	26·51	39·34	47·27	51·81	55·76	59·34	63·69
50	29·71	34·76	49·33	58·16	63·17	67·50	71·42	76·15
60	37·48	43·19	59·33	68·97	74·40	79·08	83·30	88·38
70	45·44	51·74	69·33	79·71	85·53	90·53	95·02	100·43
80	53·54	60·39	79·33	90·41	96·58	101·88	106·63	112·33
90	61·75	69·13	89·33	101·05	107·57	113·15	118·14	124·12
100	70·06	77·93	99·33	111·67	118·50	124·34	129·56	135·81

The tabulated value χ^2_P is such that, if the random variable X has the χ^2-distribution with ν degrees of freedom, the probability that $X \geqslant \chi^2_P$ is P.

For values of ν greater than 100, $\sqrt{2\chi^2}$ is approximately normally distributed with mean $\sqrt{2\nu - 1}$ and unit variance.

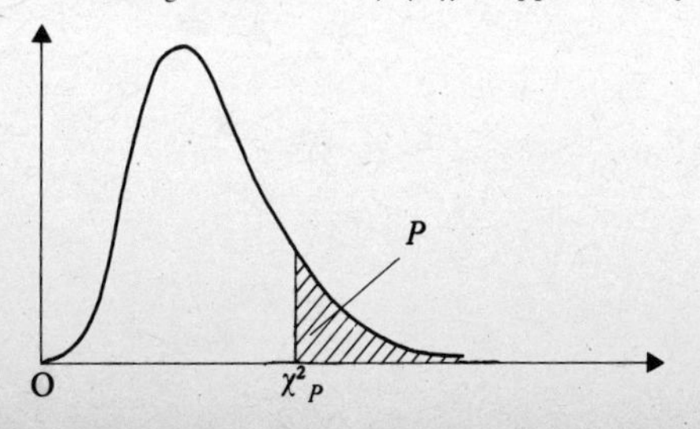

Table 33(a). Upper 5 per cent points of the F-distribution

ν_2 \ ν_1	1	2	3	4	5	6	7	8	9	10	12	15	20	24	30	∞
1	161·45	199·50	215·71	224·58	230·16	233·99	236·77	238·88	240·54	241·88	243·91	245·95	248·01	249·05	250·10	254·3
2	18·51	19·00	19·16	19·25	19·30	19·33	19·35	19·37	19·38	19·40	19·41	19·43	19·45	19·45	19·46	19·5
3	10·13	9·55	9·28	9·12	9·01	8·94	8·89	8·85	8·81	8·79	8·74	8·70	8·66	8·64	8·62	8·53
4	7·71	6·94	6·59	6·39	6·26	6·16	6·09	6·04	6·00	5·96	5·91	5·86	5·80	5·77	5·75	5·63
5	6·61	5·79	5·41	5·19	5·05	4·95	4·88	4·82	4·77	4·74	4·68	4·62	4·56	4·53	4·50	4·36
6	5·99	5·14	4·76	4·53	4·39	4·28	4·21	4·15	4·10	4·06	4·00	3·94	3·87	3·84	3·81	3·67
7	5·59	4·74	4·35	4·12	3·97	3·87	3·79	3·73	3·68	3·64	3·57	3·51	3·44	3·41	3·38	3·23
8	5·32	4·46	4·07	3·84	3·69	3·58	3·50	3·44	3·39	3·35	3·28	3·22	3·15	3·12	3·08	2·93
9	5·12	4·26	3·86	3·63	3·48	3·37	3·29	3·23	3·18	3·14	3·07	3·01	2·94	2·90	2·86	2·71
10	4·96	4·10	3·71	3·48	3·33	3·22	3·14	3·07	3·02	2·98	2·91	2·84	2·77	2·74	2·70	2·54
11	4·84	3·98	3·59	3·36	3·20	3·09	3·01	2·95	2·90	2·85	2·79	2·72	2·65	2·61	2·57	2·40
12	4·75	3·89	3·49	3·26	3·11	3·00	2·91	2·85	2·80	2·75	2·69	2·62	2·54	2·51	2·47	2·30
13	4·67	3·81	3·41	3·18	3·03	2·92	2·83	2·77	2·71	2·67	2·60	2·53	2·46	2·42	2·38	2·21
14	4·60	3·74	3·34	3·11	2·96	2·85	2·76	2·70	2·65	2·60	2·53	2·46	2·39	2·35	2·31	2·13
15	4·54	3·68	3·29	3·06	2·90	2·79	2·71	2·64	2·59	2·54	2·48	2·40	2·33	2·29	2·25	2·07
16	4·49	3·63	3·24	3·01	2·85	2·74	2·66	2·59	2·54	2·49	2·42	2·35	2·28	2·24	2·19	2·01
17	4·45	3·59	3·20	2·96	2·81	2·70	2·61	2·55	2·49	2·45	2·38	2·31	2·23	2·19	2·15	1·96
18	4·41	3·55	3·16	2·93	2·77	2·66	2·58	2·51	2·46	2·41	2·34	2·27	2·19	2·15	2·11	1·92
19	4·38	3·52	3·13	2·90	2·74	2·63	2·54	2·48	2·42	2·38	2·31	2·23	2·16	2·11	2·07	1·88
20	4·35	3·49	3·10	2·87	2·71	2·60	2·51	2·45	2·39	2·35	2·28	2·20	2·12	2·08	2·04	1·84
21	4·32	3·47	3·07	2·84	2·68	2·57	2·49	2·42	2·37	2·32	2·25	2·18	2·10	2·05	2·01	1·81
22	4·30	3·44	3·05	2·82	2·66	2·55	2·46	2·40	2·34	2·30	2·23	2·15	2·07	2·03	1·98	1·78
23	4·28	3·42	3·03	2·80	2·64	2·53	2·44	2·37	2·32	2·27	2·20	2·13	2·05	2·01	1·96	1·76
24	4·26	3·40	3·01	2·78	2·62	2·51	2·42	2·36	2·30	2·25	2·18	2·11	2·03	1·98	1·94	1·73
25	4·24	3·39	2·99	2·76	2·60	2·49	2·40	2·34	2·28	2·24	2·16	2·09	2·01	1·96	1·92	1·71
26	4·23	3·37	2·98	2·74	2·59	2·47	2·39	2·32	2·27	2·22	2·15	2·07	1·99	1·95	1·90	1·69
27	4·21	3·35	2·96	2·73	2·57	2·46	2·37	2·31	2·25	2·20	2·13	2·06	1·97	1·93	1·88	1·67
28	4·20	3·34	2·95	2·71	2·56	2·45	2·36	2·29	2·24	2·19	2·12	2·04	1·96	1·91	1·87	1·65
29	4·18	3·33	2·93	2·70	2·55	2·43	2·35	2·28	2·22	2·18	2·10	2·03	1·94	1·90	1·85	1·64
30	4·17	3·32	2·92	2·69	2·53	2·42	2·33	2·27	2·21	2·16	2·09	2·01	1·93	1·89	1·84	1·62
40	4·08	3·23	2·84	2·61	2·45	2·34	2·25	2·18	2·12	2·08	2·00	1·92	1·84	1·79	1·74	1·51
60	4·00	3·15	2·76	2·53	2·37	2·25	2·17	2·10	2·04	1·99	1·92	1·84	1·75	1·70	1·65	1·39
120	3·92	3·07	2·68	2·45	2·29	2·18	2·09	2·02	1·96	1·91	1·83	1·75	1·66	1·61	1·55	1·25
∞	3·84	3·00	2·60	2·37	2·21	2·10	2·01	1·94	1·88	1·83	1·75	1·67	1·57	1·52	1·46	1·00

In Table 33 the function tabulated is F_P with (a) $P = 0{\cdot}05$; (b) $P = 0{\cdot}025$; and (c) $P = 0{\cdot}01$.

If the random variable X has the F-distribution with ν_1 and ν_2 degrees of freedom, the probability that $X \geqslant F_P$ is P.

For $\nu_1 > 12$ or $\nu_2 > 40$ interpolation should be linear in $\frac{60}{\nu_1}$ and $\frac{120}{\nu_2}$ (see p.7).

Table 33(b). Upper $2\frac{1}{2}$ per cent points of the F-distribution

ν_2 \ ν_1	1	2	3	4	5	6	7	8	9	10	12	15	20	24	30	∞
1	647·79	799·50	864·16	899·58	921·85	937·11	948·22	956·66	963·28	968·63	976·71	984·87	993·10	997·25	1001·4	1018
2	38·51	39·00	39·17	39·25	39·30	39·33	39·36	39·37	39·39	39·40	39·41	39·43	39·45	39·46	39·46	39·5
3	17·44	16·04	15·44	15·10	14·88	14·73	14·62	14·54	14·47	14·42	14·34	14·25	14·17	14·12	14·08	13·9
4	12·22	10·65	9·98	9·60	9·36	9·20	9·07	8·98	8·90	8·84	8·75	8·66	8·56	8·51	8·46	8·26
5	10·01	8·43	7·76	7·39	7·15	6·98	6·85	6·76	6·68	6·62	6·52	6·43	6·33	6·28	6·23	6·02
6	8·81	7·26	6·60	6·23	5·99	5·82	5·70	5·60	5·52	5·46	5·37	5·27	5·17	5·12	5·07	4·85
7	8·07	6·54	5·89	5·52	5·29	5·12	4·99	4·90	4·82	4·76	4·67	4·57	4·47	4·41	4·36	4·14
8	7·57	6·06	5·42	5·05	4·82	4·65	4·53	4·43	4·36	4·30	4·20	4·10	4·00	3·95	3·89	3·67
9	7·21	5·71	5·08	4·72	4·48	4·32	4·20	4·10	4·03	3·96	3·87	3·77	3·67	3·61	3·56	3·33
10	6·94	5·46	4·83	4·47	4·24	4·07	3·95	3·85	3·78	3·72	3·62	3·52	3·42	3·37	3·31	3·08
11	6·72	5·26	4·63	4·28	4·04	3·88	3·76	3·66	3·59	3·53	3·43	3·33	3·23	3·17	3·12	2·88
12	6·55	5·10	4·47	4·12	3·89	3·73	3·61	3·51	3·44	3·37	3·28	3·18	3·07	3·02	2·96	2·72
13	6·41	4·97	4·35	4·00	3·77	3·60	3·48	3·39	3·31	3·25	3·15	3·05	2·95	2·89	2·84	2·60
14	6·30	4·86	4·24	3·89	3·66	3·50	3·38	3·29	3·21	3·15	3·05	2·95	2·84	2·79	2·73	2·49
15	6·20	4·77	4·15	3·80	3·58	3·41	3·29	3·20	3·12	3·06	2·96	2·86	2·76	2·70	2·64	2·40
16	6·12	4·69	4·08	3·73	3·50	3·34	3·22	3·12	3·05	2·99	2·89	2·79	2·68	2·63	2·57	2·32
17	6·04	4·62	4·01	3·66	3·44	3·28	3·16	3·06	2·98	2·92	2·82	2·72	2·62	2·56	2·50	2·25
18	5·98	4·56	3·95	3·61	3·38	3·22	3·10	3·01	2·93	2·87	2·77	2·67	2·56	2·50	2·44	2·19
19	5·92	4·51	3·90	3·56	3·33	3·17	3·05	2·96	2·88	2·82	2·72	2·62	2·51	2·45	2·39	2·13
20	5·87	4·46	3·86	3·51	3·29	3·13	3·01	2·91	2·84	2·77	2·68	2·57	2·46	2·41	2·35	2·09
21	5·83	4·42	3·82	3·48	3·25	3·09	2·97	2·87	2·80	2·73	2·64	2·53	2·42	2·37	2·31	2·04
22	5·79	4·38	3·78	3·44	3·22	3·05	2·93	2·84	2·76	2·70	2·60	2·50	2·39	2·33	2·27	2·00
23	5·75	4·35	3·75	3·41	3·18	3·02	2·90	2·81	2·73	2·67	2·57	2·47	2·36	2·30	2·24	1·97
24	5·72	4·32	3·72	3·38	3·15	2·99	2·87	2·78	2·70	2·64	2·54	2·44	2·33	2·27	2·21	1·94
25	5·69	4·29	3·69	3·35	3·13	2·97	2·85	2·75	2·68	2·61	2·51	2·41	2·30	2·24	2·18	1·91
26	5·66	4·27	3·67	3·33	3·10	2·94	2·82	2·73	2·65	2·59	2·49	2·39	2·28	2·22	2·16	1·88
27	5·63	4·24	3·65	3·31	3·08	2·92	2·80	2·71	2·63	2·57	2·47	2·36	2·25	2·19	2·13	1·85
28	5·61	4·22	3·63	3·29	3·06	2·90	2·78	2·69	2·61	2·55	2·45	2·34	2·23	2·17	2·11	1·83
29	5·59	4·20	3·61	3·27	3·04	2·88	2·76	2·67	2·59	2·53	2·43	2·32	2·21	2·15	2·09	1·81
30	5·57	4·18	3·59	3·25	3·03	2·87	2·75	2·65	2·57	2·51	2·41	2·31	2·20	2·14	2·07	1·79
40	5·42	4·05	3·46	3·13	2·90	2·74	2·62	2·53	2·45	2·39	2·29	2·18	2·07	2·01	1·94	1·64
60	5·29	3·93	3·34	3·01	2·79	2·63	2·51	2·41	2·33	2·27	2·17	2·06	1·94	1·88	1·82	1·48
120	5·15	3·80	3·23	2·89	2·67	2·52	2·39	2·30	2·22	2·16	2·05	1·94	1·82	1·76	1·69	1·31
∞	5·02	3·69	3·12	2·79	2·57	2·41	2·29	2·19	2·11	2·05	1·94	1·83	1·71	1·64	1·57	1·00

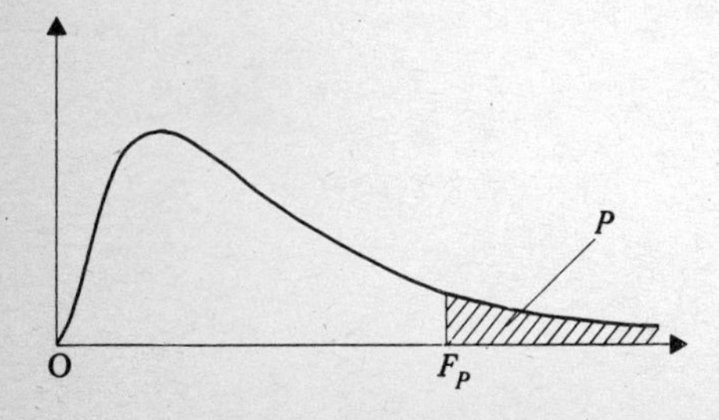

Table 33(c). Upper 1 per cent points of the *F*-distribution

ν_2 \ ν_1	1	2	3	4	5	6	7	8	9	10	12	15	20	24	30	∞
1	4052·2	4999·5	5403·4	5624·6	5763·6	5859·0	5928·4	5981·1	6022·5	6055·8	6106·3	6157·3	6208·7	6234·6	6260·6	6366
2	98·50	99·00	99·17	99·25	99·30	99·33	99·36	99·37	99·39	99·40	99·42	99·43	99·45	99·46	99·47	99·5
3	34·12	30·82	29·46	28·71	28·24	27·91	27·67	27·49	27·35	27·23	27·05	26·87	26·69	26·60	26·50	26·1
4	21·20	18·00	16·69	15·98	15·52	15·21	14·98	14·80	14·66	14·55	14·37	14·20	14·02	13·93	13·84	13·5
5	16·26	13·27	12·06	11·39	10·97	10·67	10·46	10·29	10·16	10·05	9·89	9·72	9·55	9·47	9·38	9·02
6	13·75	10·92	9·78	9·15	8·75	8·47	8·26	8·10	7·98	7·87	7·72	7·56	7·40	7·31	7·23	6·88
7	12·25	9·55	8·45	7·85	7·46	7·19	6·99	6·84	6·72	6·62	6·47	6·31	6·16	6·07	5·99	5·65
8	11·26	8·65	7·59	7·01	6·63	6·37	6·18	6·03	5·91	5·81	5·67	5·52	5·36	5·28	5·20	4·86
9	10·56	8·02	6·99	6·42	6·06	5·80	5·61	5·47	5·35	5·26	5·11	4·96	4·81	4·73	4·65	4·31
10	10·04	7·56	6·55	5·99	5·64	5·39	5·20	5·06	4·94	4·85	4·71	4·56	4·41	4·33	4·25	3·91
11	9·65	7·21	6·22	5·67	5·32	5·07	4·89	4·74	4·63	4·54	4·40	4·25	4·10	4·02	3·94	3·60
12	9·33	6·93	5·95	5·41	5·06	4·82	4·64	4·50	4·39	4·30	4·16	4·01	3·86	3·78	3·70	3·36
13	9·07	6·70	5·74	5·21	4·86	4·62	4·44	4·30	4·19	4·10	3·96	3·82	3·66	3·59	3·51	3·17
14	8·86	6·51	5·56	5·04	4·69	4·46	4·28	4·14	4·03	3·94	3·80	3·66	3·51	3·43	3·35	3·00
15	8·68	6·36	5·42	4·89	4·56	4·32	4·14	4·00	3·89	3·80	3·67	3·52	3·37	3·29	3·21	2·87
16	8·53	6·23	5·29	4·77	4·44	4·20	4·03	3·89	3·78	3·69	3·55	3·41	3·26	3·18	3·10	2·75
17	8·40	6·11	5·18	4·67	4·34	4·10	3·93	3·79	3·68	3·59	3·46	3·31	3·16	3·08	3·00	2·65
18	8·29	6·01	5·09	4·58	4·25	4·01	3·84	3·71	3·60	3·51	3·37	3·23	3·08	3·00	2·92	2·57
19	8·18	5·93	5·01	4·50	4·17	3·94	3·77	3·63	3·52	3·43	3·30	3·15	3·00	2·92	2·84	2·49
20	8·10	5·85	4·94	4·43	4·10	3·87	3·70	3·56	3·46	3·37	3·23	3·09	2·94	2·86	2·78	2·42
21	8·02	5·78	4·87	4·37	4·04	3·81	3·64	3·51	3·40	3·31	3·17	3·03	2·88	2·80	2·72	2·36
22	7·95	5·72	4·82	4·31	3·99	3·76	3·59	3·45	3·35	3·26	3·12	2·98	2·83	2·75	2·67	2·31
23	7·88	5·66	4·76	4·26	3·94	3·71	3·54	3·41	3·30	3·21	3·07	2·93	2·78	2·70	2·62	2·26
24	7·82	5·61	4·72	4·22	3·90	3·67	3·50	3·36	3·26	3·17	3·03	2·89	2·74	2·66	2·58	2·21
25	7·77	5·57	4·68	4·18	3·85	3·63	3·46	3·32	3·22	3·13	2·99	2·85	2·70	2·62	2·54	2·17
26	7·72	5·53	4·64	4·14	3·82	3·59	3·42	3·29	3·18	3·09	2·96	2·81	2·66	2·58	2·50	2·13
27	7·68	5·49	4·60	4·11	3·78	3·56	3·39	3·26	3·15	3·06	2·93	2·78	2·63	2·55	2·47	2·10
28	7·64	5·45	4·57	4·07	3·75	3·53	3·36	3·23	3·12	3·03	2·90	2·75	2·60	2·52	2·44	2·06
29	7·60	5·42	4·54	4·04	3·73	3·50	3·33	3·20	3·09	3·00	2·87	2·73	2·57	2·49	2·41	2·03
30	7·56	5·39	4·51	4·02	3·70	3·47	3·30	3·17	3·07	2·98	2·84	2·70	2·55	2·47	2·39	2·01
40	7·31	5·18	4·31	3·83	3·51	3·29	3·12	2·99	2·89	2·80	2·66	2·52	2·37	2·29	2·20	1·80
60	7·08	4·98	4·13	3·65	3·34	3·12	2·95	2·82	2·72	2·63	2·50	2·35	2·20	2·12	2·03	1·60
120	6·85	4·79	3·95	3·48	3·17	2·96	2·79	2·66	2·56	2·47	2·34	2·19	2·03	1·95	1·86	1·38
∞	6·63	4·61	3·78	3·32	3·02	2·80	2·64	2·51	2·41	2·32	2·18	2·04	1·88	1·79	1·70	1·00

Table 34. Random numbers

19	21	17	33	36	80	58	60	86	81	28	01	24	88	81	34	32	14	01	56	03	77	64	51	50
94	52	04	44	51	07	03	23	65	61	41	31	12	84	21	95	90	89	12	80	74	62	78	63	59
70	98	60	38	17	40	25	16	13	10	25	94	09	24	11	34	79	68	54	16	00	99	39	94	87
65	24	97	96	77	03	15	50	92	32	96	78	41	71	26	50	35	08	64	69	41	30	06	27	15
82	10	20	30	98	01	78	50	06	53	39	43	84	36	60	02	40	60	84	04	24	54	08	00	00
91	60	09	46	35	35	39	28	17	37	01	50	50	49	67	91	09	70	20	11	26	64	23	85	40
20	55	98	53	61	20	09	34	60	00	83	30	49	66	24	62	54	14	17	22	79	67	69	89	70
53	30	57	95	44	99	93	78	77	27	32	21	01	94	38	58	25	07	72	65	02	99	80	29	73
57	10	88	64	98	14	15	86	06	97	41	67	31	80	87	46	08	81	12	38	82	13	57	90	35
08	27	01	19	29	92	04	03	73	90	71	19	05	89	52	98	70	24	16	38	95	72	52	27	98
90	11	92	32	06	75	63	46	00	53	90	72	42	90	80	69	42	36	68	15	11	89	61	86	07
45	12	46	96	07	17	07	86	17	47	15	89	16	99	04	79	58	96	81	37	19	00	61	90	45
83	08	40	25	89	37	66	06	38	82	99	02	53	48	31	92	04	82	36	71	68	89	57	37	95
04	68	53	10	35	93	82	81	61	59	05	01	55	48	00	76	53	42	29	74	13	58	90	18	01
61	34	90	45	38	89	31	82	76	93	02	67	43	43	68	24	72	04	06	82	20	94	03	73	92
14	08	26	50	20	49	95	60	13	36	41	68	50	17	58	49	24	25	21	22	01	03	06	03	78
82	61	55	34	77	58	01	46	22	29	72	64	03	20	42	73	52	11	41	66	45	85	00	23	72
50	94	27	86	33	16	58	81	92	75	62	25	82	07	73	67	60	19	30	65	69	00	20	39	85
76	38	17	74	55	81	21	80	25	20	22	90	08	01	30	61	55	49	89	01	26	93	97	87	32
05	64	53	50	63	85	93	22	24	10	31	35	75	47	90	39	70	79	43	48	11	96	98	97	55
76	59	18	37	50	46	13	77	49	89	39	93	13	30	68	35	15	54	94	86	28	15	60	45	56
31	94	58	79	60	04	85	24	14	11	63	10	54	41	16	95	25	00	40	46	59	21	16	72	70
08	64	88	98	22	04	17	03	83	65	23	84	26	19	17	57	45	30	34	95	61	43	02	01	54
32	51	10	79	99	18	92	07	70	45	44	29	98	50	57	51	39	51	74	57	24	20	70	27	30
79	34	85	61	94	58	14	58	86	45	84	86	74	15	94	28	14	88	49	85	89	94	92	66	89
61	97	30	36	60	32	98	87	06	89	17	79	46	13	40	58	31	13	25	69	23	94	98	56	26
92	03	26	01	27	34	06	62	81	49	22	35	21	29	07	53	78	88	66	48	57	64	90	78	87
74	60	97	10	72	63	95	85	83	36	67	81	44	05	98	12	62	63	07	54	75	89	54	21	94
98	66	87	60	74	25	63	45	69	13	88	25	44	16	47	05	39	86	94	63	49	77	83	13	82
65	24	87	20	78	58	63	48	86	78	21	76	46	79	40	45	66	68	46	64	35	71	44	30	81
82	00	29	69	16	94	13	87	47	39	99	12	20	39	04	46	05	29	72	77	60	24	33	74	24
79	10	05	59	38	23	21	11	01	11	17	11	59	05	77	94	20	20	10	63	85	52	26	43	78
30	92	37	17	10	70	25	70	55	96	42	31	00	24	49	31	21	15	00	25	99	74	47	80	84
90	51	35	09	66	78	98	17	03	91	45	93	21	35	35	21	68	16	65	89	94	91	50	88	55
94	47	47	93	56	16	09	89	58	06	79	25	21	41	90	88	72	23	98	87	15	55	35	83	86
65	23	66	29	48	19	96	82	20	71	49	89	89	61	40	80	26	45	75	80	56	77	56	31	38
80	50	20	41	92	84	28	73	25	89	50	66	46	38	46	71	59	06	72	20	71	50	32	79	42
01	31	50	46	32	50	20	28	91	48	41	55	61	15	84	35	91	61	39	79	25	01	63	25	11
81	52	57	66	70	88	71	42	86	81	56	54	08	49	63	85	54	36	97	15	86	19	27	93	73
19	50	04	17	20	79	21	42	00	79	42	05	32	98	44	02	29	41	13	06	78	53	76	50	98
25	81	27	70	90	45	19	89	81	62	13	78	26	05	96	99	09	25	01	88	65	40	56	32	27
80	85	99	42	20	92	30	90	19	98	45	09	02	48	15	13	41	58	69	89	01	67	73	90	92
41	10	73	35	61	04	37	64	00	72	78	90	96	10	42	04	09	87	33	04	21	89	26	31	12
00	46	50	08	58	22	77	48	07	30	07	09	88	05	15	09	97	04	04	76	10	31	42	47	92
58	13	70	24	54	15	65	72	49	57	48	40	10	29	40	92	82	82	63	72	31	07	15	81	92
32	01	39	71	47	69	72	57	88	67	73	32	97	49	35	69	27	64	60	01	04	18	13	88	38
17	04	88	47	88	12	53	10	17	73	43	55	13	45	86	61	23	98	79	27	03	23	23	13	12
33	93	50	79	44	98	45	61	19	22	96	17	42	41	00	00	30	78	56	97	06	52	73	43	81
47	63	34	93	94	38	67	32	22	81	68	09	67	65	99	38	46	21	66	62	81	95	90	33	58
82	16	19	25	21	10	71	25	88	39	18	54	63	29	20	89	22	09	04	93	73	72	52	23	27

Table 35. The Wilcoxon signed rank test: critical and quasi-critical values of R_+.

n \ P	·05		·025		·01		·005	
5	0	0·0313						
	1	0·0625						
6	2	0·0469	0	0·0156				
	3	0·0781	1	0·0313				
7	3	0·0391	2	0·0234	0	0·0078		
	4	0·0547	3	0·0391	1	0·0156		
8	5	0·0391	3	0·0195	1	0·0078	0	0·0039
	6	0·0547	4	0·0273	2	0·0117	1	0·0078
9	8	0·0488	5	0·0195	3	0·0098	1	0·0039
	9	0·0645	6	0·0273	4	0·0137	2	0·0059
10	10	0·0420	8	0·0244	5	0·0098	3	0·0049
	11	0·0527	9	0·0322	6	0·0137	4	0·0068
11	13	0·0415	10	0·0210	7	0·0093	5	0·0049
	14	0·0508	11	0·0269	8	0·0122	6	0·0068
12	17	0·0461	13	0·0212	9	0·0081	7	0·0046
	18	0·0549	14	0·0261	10	0·0105	8	0·0061
13	21	0·0471	17	0·0239	12	0·0085	9	0·0040
	22	0·0549	18	0·0287	13	0·0107	10	0·0052
14	25	0·0453	21	0·0247	15	0·0083	12	0·0043
	26	0·0520	22	0·0290	16	0·0101	13	0·0054
15	30	0·0473	25	0·0240	19	0·0090	15	0·0042
	31	0·0535	26	0·0277	20	0·0108	16	0·0051
16	35	0·0467	29	0·0222	23	0·0091	19	0·0046
	36	0·0523	30	0·0253	24	0·0107	20	0·0055
17	41	0·0492	34	0·0224	27	0·0087	23	0·0047
	42	0·0544	35	0·0253	28	0·0101	24	0·0055
18	47	0·0494	40	0·0241	32	0·0091	27	0·0045
	48	0·0542	41	0·0269	33	0·0104	28	0·0052
19	53	0·0478	46	0·0247	37	0·0090	32	0·0047
	54	0·0521	47	0·0273	38	0·0102	33	0·0054
20	60	0·0487	52	0·0242	43	0·0096	37	0·0047
	61	0·0527	53	0·0266	44	0·0107	38	0·0053

The Wilcoxon signed rank test

n \ P	·05		·025		·01		·005	
21	67	0·0479	58	0·0230	49	0·0097	42	0·0045
	68	0·0516	59	0·0251	50	0·0108	43	0·0051
22	75	0·0492	65	0·0231	55	0·0095	48	0·0046
	76	0·0527	66	0·0250	56	0·0104	49	0·0052
23	83	0·0490	73	0·0242	62	0·0098	54	0·0046
	84	0·0523	74	0·0261	63	0·0107	55	0·0051
24	91	0·0475	81	0·0245	69	0·0097	61	0·0048
	92	0·0505	82	0·0263	70	0·0106	62	0·0053
25	100	0·0479	89	0·0241	76	0·0094	68	0·0048
	101	0·0507	90	0·0258	77	0·0101	69	0·0053
26	110	0·0497	98	0·0247	84	0·0095	75	0·0047
	111	0·0524	99	0·0263	85	0·0102	76	0·0051
27	119	0·0477	107	0·0246	92	0·0093	83	0·0048
	120	0·0502	108	0·0260	93	0·0100	84	0·0052
28	130	0·0496	116	0·0239	101	0·0096	91	0·0048
	131	0·0521	117	0·0252	102	0·0102	92	0·0051
29	140	0·0482	126	0·0240	110	0·0095	100	0·0049
	141	0·0504	127	0·0253	111	0·0101	101	0·0053
30	151	0·0481	137	0·0249	120	0·0098	109	0·0050
	152	0·0502	138	0·0261	121	0·0104	110	0·0053

R_+ and R_- are the sums of like-signed ranks derived from n matched pairs. Wilcoxon's T is the smaller of R_+ and R_-. The table refers to the distribution of R_+ when the two population distributions are identical.

For each n and P the table gives two integers, R_1 and R_2, followed by the probabilities that $R_+ \leqslant R_1$ and that $R_+ \leqslant R_2$. These probabilities are the nearest to P that exist for integral values of R_+.

For $n > 30$, R_+ is approximately normally distributed with mean $\frac{1}{4}n(n+1)$ and variance $\frac{1}{24}n(n+1)(2n+1)$.

Table 36. SI units

The Système International, or SI, is a rationalized system of units based on the metric system. It is an internationally agreed system which incorporates the four base units of the old MKSA system, and defines three new ones. From these, and the two supplementary units, all the other units can be derived.

Base units

Quantity	Name	Symbol
length	metre	m
mass	kilogram	kg
time	second	s
electric current	ampere	A
thermodynamic temperature	kelvin	K
amount of substance	mole	mol
luminous intensity	candela	cd

Supplementary units

Quantity	Name	Symbol
plane angle	radian	rad
solid angle	steradian	sr

Prefixes for units

Prefix	Symbol	Factor	Prefix	Symbol	Factor
exa	E	10^{18}	deci	d	10^{-1}
peta	P	10^{15}	centi	c	10^{-2}
tera	T	10^{12}	milli	m	10^{-3}
giga	G	10^{9}	micro	μ	10^{-6}
mega	M	10^{6}	nano	n	10^{-9}
kilo	k	10^{3}	pico	p	10^{-12}
hecto	h	10^{2}	femto	f	10^{-15}
deca	da	10^{1}	atto	a	10^{-18}

Derived units having special names and symbols

Quantity	Name	Symbol	Expressed in terms of SI units
activity (of a radioactive source)	becquerel	Bq	s^{-1}
quantity of electricity, electric charge	coulomb	C	A s
electric capacitance	farad	F	$c\ V^{-1}$
absorbed dose (of ionizing radiation)	gray	Gy	$J\ kg^{-1}$
inductance	henry	H	$Wb\ A^{-1}$
frequency	hertz	Hz	s^{-1}
energy, work, quantity of heat	joule	J	N m
luminous flux	lumen	lm	cd sr
illuminance	lux	lx	$lm\ m^{-2}$
force	newton	N	$m\ kg\ s^{-2}$
electric resistance	ohm	Ω	$V\ A^{-1}$
pressure	pascal	Pa	$N\ m^{-2}$
electric conductance	siemens	S	Ω^{-1}
magnetic flux density	tesla	T	$Wb\ m^{-2}$
electric potential difference	volt	V	$W\ A^{-1}$
power	watt	W	$J\ s^{-1}$
magnetic flux	weber	Wb	V s

Table 37. Conversion factors

Length

1 inch (in)	2·54 cm
1 foot (ft)	0·3048 m
1 yard	0·9144 m
1 mile	1·609 km
1 nautical mile (Int)	1·852 km
1 light year	$9{\cdot}461 \times 10^{15}$ m
1 astronomical unit	$1{\cdot}496 \times 10^{11}$ m
1 parsec	3·263 light years

Area

1 hectare	10^4 m^2
1 square inch	6·45 cm^2
1 square foot	0·0929 m^2
1 square yard	0·836 m^2
1 acre	0·4047 ha
1 square mile	2·59 km^2

Volume

1 millilitre (ml)	1 cm^3
1 litre (l)	1 dm^3
1 cubic inch	16·4 cm^3
1 cubic foot	0·0283 m^3
1 cubic yard	0·765 m^3
1 pint	0·568 l
1 gallon (UK)	4·546 l
1 gallon (US)	3·785 l

Speed

1 km per hour	0·2778 m s^{-1}
1 mile per hour	0·4470 m s^{-1}
1 knot (Int)	0·5144 m s^{-1}
1 rev per minute	0·1047 rad s^{-1}

Mass and density

1 gram (g)	10^{-3} kg
1 tonne (t)	10^3 kg
1 ounce (oz)	28·35 g
1 pound (lb)	0·4536 kg
1 ton	1·016 t
1 lb in^{-3}	27·7 g cm^{-3}
1 lb ft^{-3}	16·02 kg m^{-3}

Force

1 poundal (pdl)	0·1383 N
1 pound force (lbf)	4·448 N

Energy and Power

1 foot poundal	0·0421 J
1 calorie I.T.	4·1868 J
1 B.t.u.	1·055 J
1 horsepower	745·7 W

Pressure

1 bar	10^5 Pa
1 atmosphere	1·0133 bar
1 pound force per square inch	6·895 kPa

Table 38. Proportional parts for tenths

Δ \ p	·1	·2	·3	·4	·5	·6	·7	·8	·9	Δ \ p	·1	·2	·3	·4	·5	·6	·7	·8	·9
1	0	0	0	0	1	1	1	1	1	51	5	10	15	20	26	31	36	41	46
2	0	0	1	1	1	1	1	2	2	52	5	10	16	21	26	31	36	42	47
3	0	1	1	1	2	2	2	2	3	53	5	11	16	21	27	32	37	42	48
4	0	1	1	2	2	2	3	3	4	54	5	11	16	22	27	32	38	43	49
5	1	1	2	2	3	3	4	4	5	55	6	11	17	22	28	33	39	44	50
6	1	1	2	2	3	4	4	5	5	56	6	11	17	22	28	34	39	45	50
7	1	1	2	3	4	4	5	6	6	57	6	11	17	23	29	34	40	46	51
8	1	2	2	3	4	5	6	6	7	58	6	12	17	23	29	35	41	46	52
9	1	2	3	4	5	5	6	7	8	59	6	12	18	24	30	35	41	47	53
10	1	2	3	4	5	6	7	8	9	60	6	12	18	24	30	36	42	48	54
11	1	2	3	4	6	7	8	9	10	61	6	12	18	24	31	37	43	49	55
12	1	2	4	5	6	7	8	10	11	62	6	12	19	25	31	37	43	50	56
13	1	3	4	5	7	8	9	10	12	63	6	13	19	25	32	38	44	50	57
14	1	3	4	6	7	8	10	11	13	64	6	13	19	26	32	38	45	51	58
15	2	3	5	6	8	9	11	12	14	65	7	13	20	26	33	39	46	52	59
16	2	3	5	6	8	10	11	13	14	66	7	13	20	26	33	40	46	53	59
17	2	3	5	7	9	10	12	14	15	67	7	13	20	27	34	40	47	54	60
18	2	4	5	7	9	11	13	14	16	68	7	14	20	27	34	41	48	54	61
19	2	4	6	8	10	11	13	15	17	69	7	14	21	28	35	41	48	55	62
20	2	4	6	8	10	12	14	16	18	70	7	14	21	28	35	42	49	56	63
21	2	4	6	8	11	13	15	17	19	71	7	14	21	28	36	43	50	57	64
22	2	4	7	9	11	13	15	18	20	72	7	14	22	29	36	43	50	58	65
23	2	5	7	9	12	14	16	18	21	73	7	15	22	29	37	44	51	58	66
24	2	5	7	10	12	14	17	19	22	74	7	15	22	30	37	44	52	59	67
25	3	5	8	10	13	15	18	20	23	75	8	15	23	30	38	45	53	60	68
26	3	5	8	10	13	16	18	21	23	76	8	15	23	30	38	46	53	61	68
27	3	5	8	11	14	16	19	22	24	77	8	15	23	31	39	46	54	62	69
28	3	6	8	11	14	17	20	22	25	78	8	16	23	31	39	47	55	62	70
29	3	6	9	12	15	17	20	23	26	79	8	16	24	32	40	47	55	63	71
30	3	6	9	12	15	18	21	24	27	80	8	16	24	32	40	48	56	64	72
31	3	6	9	12	16	19	22	25	28	81	8	16	24	32	41	49	57	65	73
32	3	6	10	13	16	19	22	26	29	82	8	16	25	33	41	49	57	66	74
33	3	7	10	13	17	20	23	26	30	83	8	17	25	33	42	50	58	66	75
34	3	7	10	14	17	20	24	27	31	84	8	17	25	34	42	50	59	67	76
35	4	7	11	14	18	21	25	28	32	85	9	17	26	34	43	51	60	68	77
36	4	7	11	14	18	22	25	29	32	86	9	17	26	34	43	52	60	69	77
37	4	7	11	15	19	22	26	30	33	87	9	17	26	35	44	52	61	70	78
38	4	8	11	15	19	23	27	30	34	88	9	18	26	35	44	53	62	70	79
39	4	8	12	16	20	23	27	31	35	89	9	18	27	36	45	53	62	71	80
40	4	8	12	16	20	24	28	32	36	90	9	18	27	36	45	54	63	72	81
41	4	8	12	16	21	25	29	33	37	91	9	18	27	36	46	55	64	73	82
42	4	8	13	17	21	25	29	34	38	92	9	18	28	37	46	55	64	74	83
43	4	9	13	17	22	26	30	34	39	93	9	19	28	37	47	56	65	74	84
44	4	9	13	18	22	26	31	35	40	94	9	19	28	38	47	56	66	75	85
45	5	9	14	18	23	27	32	36	41	95	10	19	29	38	48	57	67	76	86
46	5	9	14	18	23	28	32	37	41	96	10	19	29	38	48	58	67	77	86
47	5	9	14	19	24	28	33	38	42	97	10	19	29	39	49	58	68	78	87
48	5	10	14	19	24	29	34	38	43	98	10	20	29	39	49	59	69	78	88
49	5	10	15	20	25	29	34	39	44	99	10	20	30	40	50	59	69	79	89
50	5	10	15	20	25	30	35	40	45	100	10	20	30	40	50	60	70	80	90

The table gives the value of $p\Delta$ rounded to the nearest integer.

MATHEMATICS BOOKS IN PENGUIN AND PELICAN

General Mathematics
CONCEPTS OF MODERN MATHEMATICS – Stewart
GEOMETRY AND THE LIBERAL ARTS – Pedoe
MATHEMATICS AND LOGIC – Kac and Ulam
MATHEMATICS AND THE IMAGINATION – Kasner and Newman
MATHEMATICS IN MANAGEMENT – Battersby
MATHEMATICS IN WESTERN CULTURE – Kline
NEWER USES OF MATHEMATICS – Ed. Lighthill
THE PSYCHOLOGY OF LEARNING MATHEMATICS – Skemp
RIDDLES IN MATHEMATICS – Northrop
MATHEMATICIAN'S DELIGHT – Sawyer
MATHEMATICS FOR TECHNOLOGY – Dobinson
VOL 1
VOL 2

Forthcoming
CATASTROPHE THEORY – Woodcock and Davis
REASON BY NUMBERS – Moore
GÖDEL, ESCHER, BACH – Hofstadter
WHAT IS THE NAME OF THIS BOOK? – Smullyan
THE AMBIDEXTROUS UNIVERSE 2nd Edition – Gardner

Recreational Mathematics
FURTHER MATHEMATICAL PUZZLES AND DIVERSIONS – Gardner
MATHEMATICAL PUZZLES AND DIVERSIONS – Gardner
MORE MATHEMATICAL PUZZLES AND DIVERSIONS – Gardner
MATHEMATICAL CARNIVAL – Gardner
THE MOSCOW PUZZLES – Kordemsky

Statistics
FACTS FROM FIGURES – Moroney
HOW TO LIE WITH STATISTICS – Huff
HOW TO TAKE A CHANCE – Huff
LADY LUCK – Weaver
STATISTICS IN ACTION – Sprent
THE USE AND ABUSE OF STATISTICS – Reichmann
STATISTICS FOR THE SOCIAL SCIENTIST – Yoemans
VOL 1: INTRODUCING STATISTICS
VOL 2: APPLIED STATISTICS

Forthcoming
STATISTICS WITHOUT TEARS – Rowntree
QUICK STATISTICS – Sprent

Calculators and Computing
A DICTIONARY OF COMPUTERS – Chandor
ELECTRONIC COMPUTERS – Hollingdale and Toothill

Forthcoming
ADVENTURES WITH YOUR POCKET CALCULATOR – Råde and Kaufman
THE CREATIVE USE OF CALCULATORS – Killingbeck
A DICTIONARY OF MICROPROCESSORS – Chandor